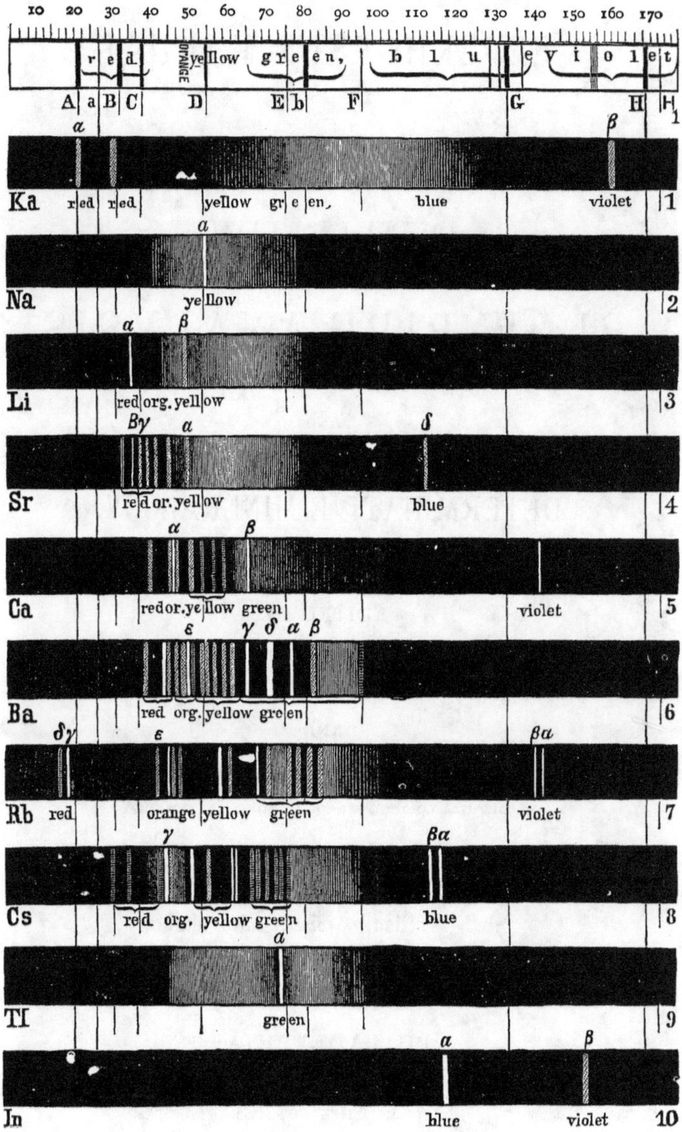

TABLE OF SPECTRA ACCORDING TO KIRCHOFF & BUNSEN.

ELDERHORST'S MANUAL

OF

QUALITATIVE

BLOW-PIPE ANALYSIS,

AND

DETERMINATIVE MINERALOGY.

EDITED BY

HENRY B. NASON, PH. D.,

RENSSELAER POLYTECHNIC INSTITUTE, TROY, N. Y.,

AND

CHARLES F. CHANDLER, PH. D.,

COLUMBIA COLLEGE SCHOOL OF MINES, NEW YORK.

Sixth Edition, Revised and Enlarged.

PHILADELPHIA:

PORTER & COATES,

No. 822 CHESTNUT STREET.

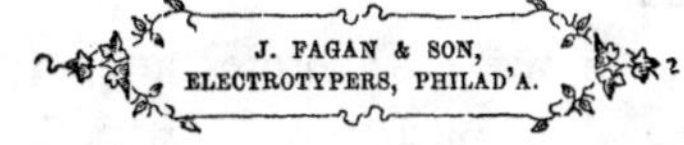

CAXTON PRESS OF SHERMAN & CO.

PREFACE

TO THE

FOURTH EDITION.

THE editors have made a complete revision of this fourth edition of Elderhorst's Manual. While many alterations and additions have been made, both in methods of determination and mineral species, they have endeavored not to make the book too large, or destroy its original plan.

The names of minerals and ores, and, in many cases, the formulæ have been made to agree with those given in the last edition of Dana's Mineralogy.

The few pages on the spectroscope and Bunsen's flame reactions, it is thought, will be found useful and convenient for the student.

It is hoped this edition will meet with the same favor, and will be found as valuable a guide, as those which have preceded it.

H. B. N.
C. F. C.

PREFACE TO THE THIRD EDITION.

THE present edition of this "Manual" is, like the preceding, designed to serve as a text-book in the instruction in Blow-pipe Analysis and Determinative Mineralogy, in the Rensselaer Polytechnic Institute.

In the first three chapters, but few alterations and additions have been made. The fourth chapter has been considerably enlarged by increasing the number of species, in the selection of which I have paid particular regard to those occurring in the American Continent; for this reason, many less important ores have found a place in the list to the exclusion of others, which, though more valuable, have not hitherto been found in America.

The fifth chapter, containing a systematic method for the discrimination of inorganic compounds, is a translation, but slightly altered, of the "*Division dichotomique pour reconnaître les minèraux*," as given in Laurent's "*Analyze au Chalumeau*." It is of no great value to the experienced analyst, but very useful for beginners, and it is on their account that I have given it a place in the Manual.

The sixth chapter is not contained in the first edition. It is hardly necessary to allege any reason for its introduction into this edition. The admirable method of Professor von Kobell for the discrimination of minerals is, almost beyond dispute, the most practical and most reliable that has ever been published. The sixth chapter is nothing but an extract from Prof. v. Kobell's treatise on this subject.

For the material of this compilation, the author is principally indebted to the works of Plattner, Berzelius, Von Kobell, Dana, and Mitchell.

The author, finally, begs to tender his thanks to his friend, Professor Chandler, of Union College, for the valuable suggestions he has received at his hands, and which he has acted upon to the best of his ability, being fully convinced that by adding the improvements recommended by his friend, the practical utility of this little Manual will be greatly increased.

WILLIAM ELDERHORST,
Professor of Chemistry in the R. P. Institute.

TROY, N. Y., April, 1860.

INTRODUCTION.

IN preparing this little Manual, it has been my principal care to adapt it to the use of the beginner. The use of the blow-pipe, though elaborately studied and extensively written on by some of the first chemists and mineralogists of the preceding and the present century, has not yet been duly appreciated. This neglect is, perhaps, owing to the rapid advancement of chemical analysis in the humid way, which furnishes, on the whole, more reliable results, and allows of an easy quantitative determination of the various constituents of a body. But it was overlooked that this mode of analysis absorbs much more time, and requires the use of an extensive set of apparatus, whereas an examination before the blow-pipe is sooner performed; requiring scarcely as many hours as an examination in the humid way requires days, and that with the aid only of a few reagents and instruments of small size. It is for this reason that a knowledge of blow-pipe operations is less valuable for the Chemist by profession than for the Mining Engineer, the Mineralogist, and the Geologist. A small portable box will hold all the necessary reagents and instruments, so that he may carry them with him on his expeditions and travels, and examine on the spot the minerals which he meets with on his explorations; an advantage which ought, truly, not to be overlooked.

For teachers who have not hitherto devoted much time to instruction in this department, a short exposition of the course which I have followed for a number of years may, perhaps, be desirable. For elementary instruction, the students are only furnished with the principal reagents, viz.: carbonate of soda, salt of phosphorus, borax, and solution of cobalt; of apparatus they want a fluid-lamp, blow-pipe with platinum point, platinum-pointed forceps, platinum wire, charcoal, and closed and open glass tubes. After having explained to them the action of the two cones of the flame, and instructed them in making beads, and conducting the processes of oxidation and reduction, I make them perform the most important operations, and study the behavior of the most commonly occurring substances, with and without fluxes. I give the substances in somewhat the following order:

Sesquioxide of iron, all the reactions given in Table II, 13.
Binoxide of manganese, Table II, 16.
Sesquioxide of chromium, Table II, 6.
Oxide of cobalt, and nickel, Table II, 7, 19.
Protoxide of copper, Table II, 8, and § 37.
Oxide of zinc, Table II, 35, and metallic zinc, §§ 25, 45.
Oxide of tin, Table II, 30, and metallic tin, § 26.
Oxide of lead, Table II, 15, and metallic lead, § 23.
Oxide of bismuth, Table II, 3, and metallic bismuth, §§ 17, 22.
Antimonous oxide, Table II, 1, and metallic antimony, §§ 16, 21.
Arsenous oxide, Table II, 2, §§ 9, 15.
Oxide of mercury, Table II, 17.
Alumina, Table I, 5, and § 44.
Magnesia, Table I, 4, and § 44.
Silica, § 39.
A sulphide, §§ 10, 14, 107.
A borate, § 60.
A chloride, §§ 65, 66.

Having performed all these operations, the student will be qualified to enter upon the analysis of substances of not too compound a character. If he meets on his way with bodies, the behavior of which before the blow-pipe he has not previously studied, he will not have any difficulty in determining their character if he follows the directions given in the second chapter. The *modus operandi* will be best understood by a few examples.

1. The substance under examination is sulphide of antimony.

Examination in a matrass: At a very high temperature, a black sublimate is obtained, becoming reddish-brown when cold. In reading over the list in § 10, we find this character belonging to sulphide of antimony.

Examination in an open glass tube: Gives sulphurous acid detected by the odor and action on blue litmus paper, and white fumes which partly condense in the tube. On examining the sublimate with a magnifying glass, it is found to be amorphous, hence must be antimonous acid (§ 16).

Examination on charcoal alone: Is completely volatilized with emission of sulphurous acid, and deposits a white volatile coating, possessing the properties of the coating of antimony (§ 21).

These few operations are quite sufficient to establish the nature of the substance under trial, since the absence of the more fixed metals is proved by the volatility of the substance on charcoal and in the open tube, and the absence of metals giving coatings by the purity of the antimony-coating. The presence of arsenic would have been betrayed by an alliaceous odor when heated on charcoal. The only substance which would have escaped detection by these operations is sulphide of mercury. In order to ascertain its presence or absence, we perform the operation given under "*Mercury*" in Chap. III.

The result giving an answer in the negative, the body was "sulphide of antimony."

2. The substance under examination is chromate of lead.

Examination in a matrass: } Fuses and changes color, but
Examination in an open tube: } gives nothing volatile.

Examination on charcoal alone: Fuses, gives small metallic globules, and deposits a coating which is lemon-yellow while hot, and sulphur-yellow when cold, indicative of lead (§ 23). It is always desirable to collect the metal to a large globule, and to study its physical properties. This end is best attained by mixing the substance with carbonate of soda and a little borax, and exposing the mixture to the reduction flame on charcoal. In this particular case, a metallic button is obtained which is soft, may be flattened by the hammer and cut by the knife, properties belonging to metallic lead.

Examination with borax and salt of phosphorus: Before proceeding with this examination it is necessary to test the substance for the presence of sulphur after the method given § 107 (unless the presence of this element was detected by the examination in the open glass tube or on charcoal alone); no sulphur being present, borax and salt of phosphorus beads are made on charcoal, and small portions of the substance added. With both fluxes nearly the same reactions are obtained; in oxidation flame dark-red while hot, and fine yellowish-green when cold; in reduction flame green, hot and cold. In order to find out what body produces such reactions, we use Table III, which leads us to sesquioxide of chromium. To corroborate the result, the substance may be fused with carbonate of soda and nitre, as described § 68.

The physical properties of the body under trial lead to the final conclusion that it must be chromate of lead.

3. The substance is an alloy of silver, copper, and lead.

Examination in a matrass: } No change.
Examination in an open tube: }

Examination on charcoal alone: Fuses and deposits a copious coating, which is lemon-yellow while hot and sulphur-yellow when cold, indicative of lead (§ 23); the coating cannot contain any oxide of bismuth, because the color would be darker in this case, but might contain oxide of zinc or oxide of antimony. The test is for the presence of the former, the coating is played upon with the oxidation flame: it is completely volatile, hence no zinc present (might also be tested with solution of cobalt, § 45); to test the coating for the presence of oxide of antimony, it is scraped off from the charcoal and dissolved in a bead of salt of phosphorus, v. § 87, or the alloy is treated with boracic acid as described under the head of "*Antimony*" in Chapter III. If the blast is continued for a long time, a faint dark-red coating is formed near the assay-piece, indicative of silver, § 27, and a dark metallic globule remains.

Examination with borax and salt of phosphorus: The globule remaining on the charcoal after volatilization of the lead, is treated with borax on charcoal in oxidation flame; the borax becomes colored. Owing to the reducing effect of the charcoal, the influence of the oxidation flame cannot be well observed on charcoal, hence the borax is removed from the metallic globule, fastened into the hook of a platina wire, and here exposed to the action of the oxidation flame; the bead is green while hot, and blue when cold. On consulting Table III we find that this reaction is produced by oxide of copper, and by a mixture of oxide of cobalt and sesquioxide of iron: to decide between the two, we now expose the bead to the action of the reduction flame; it becomes red and opaque, thus proving the presence of oxide of copper.

By the examination on charcoal, *per se*, we were led to suspect the presence of silver; in order to establish this beyond a doubt, we refer to Chapter III, "Silver;" here we find a method (§ 105) by which the presence of silver may be ascertained in compounds of all descriptions. In our case, having to deal only with lead, copper, and silver, the treatment with vitrified boracic acid and metallic lead is, of course, superfluous. We place our alloy at once on the cupel and direct the oxidation flame upon it; if, after cessation of the rotatory motion, the globule should not possess the bright lustre of silver, some pure metallic lead has to be added, in order to remove the last traces of copper. We finally obtain a bright globule exhibiting all the characteristic properties of silver.

Thus we have established the presence of lead, copper, and silver.

4. The substance under examination is copper nickel, containing arsenic, sulphur, nickel, cobalt, and iron.

Examination in a matrass: Gives a slight sublimate, consisting of octahedral crystals, pointing to the presence of arsenic (§ 11).

Examination in a glass tube open at both ends: Gives a copious crystalline sublimate of arsenous acid, and a faint odor of sulphurous acid; to establish the presence of sulphur beyond doubt, we refer to Chapter III, "Sulphur," where we find the method (§ 107) for discovering sulphur when in combination with other substances. In performing the test there described, we obtain the sulphur reaction.

Examination on charcoal alone: Gives abundant arsenical fumes, leaving a metallic globule which, even with continued blowing, does not give rise to the formation of a coating on the charcoal (absence of volatile metals).

Having removed all volatile substances, we now proceed to examine the remaining globule. On applying a magnet, we find it powerfully attracted, showing the presence of either iron, nickel, or cobalt, perhaps all of them, either alone or combined with other non-volatile metals. We add some borax to the globule and expose it to the action of the oxidation flame, then remove the borax from the globule, fasten it into the hook of a platinum wire, and here observe the color: green while hot, blue when cold as in the preceding case (example 3), but on exposing the bead to the action of the reduction flame (which is best done by placing it on charcoal and touching it with tin) it does not become brown and opaque, showing therefore the presence of a small quantity of iron with cobalt. We now add a fresh portion of borax to the metallic globule, in order to see whether it consists entirely of cobalt (that it cannot contain any considerable amount of iron, is proved by the appearance of the cobalt reaction in the first trial, iron being much more readily dissolved by borax than cobalt): the bead is violet while hot, and assumes a brownish color on cooling; by referring to Table III, we see that this effect is produced by nickel containing cobalt. Referring to Chapter III, "Nickel," we find the method to detect the presence of this metal when in combination with iron and cobalt, and also the presence of copper, if the assay should contain a small quantity of it.

By the above examples, the use of the methods given in the third chapter will be sufficiently illustrated. If the substance under examination is of a simple composition, its nature is readily ascertained by following the general method laid down in the second chapter; but if the reactions obtained clearly point to the complex nature of the body, we refer to the respective sections of Chapter III; if, for example, we suspect the presence of cobalt in a mineral consisting of arsenides, we test the substance according to § 69; if a small quantity of copper is to be discovered in a mineral, we proceed as directed in § 71, etc.

The student who is willing to devote more time to the subject than is usually allotted to it in our colleges, will do well to go carefully through all the reactions given in the second chapter, and thus familiarize himself with the colors and other properties of the various coatings, sublimates, etc., and also to perform the principal tests by which substances are discovered when in combination with others, which are fully given in the third chapter. In order to obtain characteristic reactions, it is important to experiment upon a suitable substance. For the benefit of the beginner, who would naturally be embarrassed in the choice of a body suitable for the experiment, I add a list of substances which, with few exceptions, are readily obtained, and which are sufficient to illustrate all the important reactions. After having mentioned a reaction, or described a process (in Chapters II and III), I have added a number in [ ] brackets. The number points to the substance of the list, given on p. xii, best adapted to illustrate the reaction. As each experiment requires only a very small quantity of the substance, they are most conveniently kept in small glass tubes of about an inch and a half in length and one-eighth of an inch in diameter. For the first fourteen substances no glass tubes are required, since they are the regular blow-pipe reagents. A small box containing seventy-five of the little tubes will hold the whole collection.

W. E.

COLLECTION OF SUBSTANCES,

Well adapted to Illustrate the Important Reactions of Bodies before the Blow-pipe.

1. Carbonate of soda.
2. Borax.
3. Salt of phosphorus.
4. Bisulphate of potassa.
5. Boracic acid.
6. Fluorite.
7. Nitrate of cobalt.
8. Oxalate of nickel.
9. Oxide of copper.
10. Chloride of silver.
11. Lead.
12. Iron.
13. Tin.
14. Bone-ash.
15. Chloride of potassium.
16. Bromide of potassium.
17. Iodide of potassium.
18. Chloride of sodium.
19. Chloride of ammonium.
20. Chlorate of potassa.
21. Alumina.
22. Sulphate of copper.
23. Nitrate of lead.
24. Oxide of antimony.
25. Arsenous oxide.
26. Oxide of bismuth.
27. Oxide of cadmium.
28. Sesquioxide of chromium.
29. Oxide of cobalt.
30. Protoxide of mercury.
31. Molybdic acid.
32. Oxide of silver.
33. Binoxide of tin.
34. Tungstic acid.
35. Sesquioxide of uranium.
36. Oxide of zinc.
37. Chloride of copper.
38. Arsenite of copper.
39. Subchloride of mercury.
40. Protochloride of mercury.
41. Antimony.
42. Arsenic.
43. Bismuth.
44. Cadmium.
45. Silver.
46. Zinc.
47. Alloy of mercury and tin.
48. Alloy of lead and antimony.
49. Alloy of lead and bismuth.
50. Alloy of lead and zinc.
51. Alloy of lead, copper, and silver.
52. Alloy of tin and copper.
53. Alloy of zinc and cadmium.
54. Quartz, rock crystal.
55. Gypsum.
56. Calcite.
57. Strontianite.
58. Witherite.
59. Magnesite.
60. Mica.
61. Orthoclase.
62. Albite.
63. Petalite.
64. Hematite.
65. Rutile.
66. Pyrolusite.
67. Lepidolite.
68. Apatite.
69. Franklinite.
70. Uraninite.
71. Chromite.
72. Cerusite.
73. Malachite.
74. Stibnite.
75. Iron pyrites.
76. Copper pyrites.
77. Mispickel.
78. Smaltite.
79. Cobaltite.
80. Realgar.
81. Cinnabar.
82. Niccolite.
83. Molybdenite.
84. Berthierite.
85. Bournonite.
86. Tetrahedrite.
87. Tiemannite, or Clausthalite.
88. Sulphides of arsenic and antimony (artificial).
89. Sulphides of arsenic, antimony, lead, and copper (artificial).
90. Retinite.

TABLE OF CONTENTS.

CHAPTER IV.

ABBREVIATIONS.

O.Fl. for Oxidation flame; R.Fl. for Reduction flame; S.Ph. for Salt of Phosphorus; Bx. for Borax; Sd. for Carbonate of Soda; So.Co. for Solution of Nitrate of Cobalt; Ch. for Charcoal; Ct. for Coating; Blp. for Blow-pipe; H. for Hardness; G. for Specific Gravity.

A MANUAL

OF

QUALITATIVE BLOW-PIPE ANALYSIS.

CHAPTER I.

Auxiliary Apparatus and Reagents.

§ 1. THE common blow-pipe of gas-fitters, jewellers, etc., is not very well adapted for analytical researches, as the narrow outlet becomes frequently obstructed by the moisture which is exhaled from the lungs and condenses in the tube. To avoid this inconvenience, the long cylindrical tube, *a b*, of the blow-pipe should be furnished at the extremity with a globular or cylindrical chamber, *c d*, for the reception of the condensed water. In this chamber the jet, *f g*, is inserted at a right angle to the tube. Silver is, in many respects, the best material for the construction of a blow-pipe, but has the disadvantage of becoming very hot when used for a long while, so that it becomes almost impossible to hold it with the naked fingers; next to silver stands German silver and brass. For jets, *h*, platinum is preferable to all other metals. A mouth-piece, *a*, of box-wood, horn or ivory is convenient, though not necessary. (Fig. 1.)

§ 2. Any kind of flame may be used for the blow-pipe, provided it be not too small. Some of the older chemists used common candles in preference, and it must be confessed that, in the majority of cases, the heat produced by the flame of a good sperm candle is quite sufficient. Berzelius recommended an oil lamp with a flat wick, which is now in general use as "Berzelius's Blow-pipe Lamp."* A common fluid (mixture of twelve parts

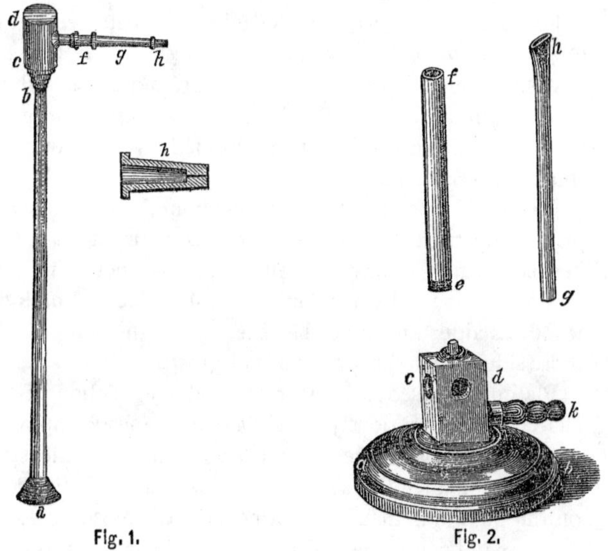

Fig. 1. Fig. 2.

of alcohol and one of turpentine, or four parts of alcohol and one of benzine) lamp, with a rather large burner, answers every purpose; it gives a very good heat, and, besides being much cleaner than an oil lamp, admits of a very quick and accurate adjustment of the size of

* Refined rapeseed or olive oil should be used in this lamp.

the flame, by means of a little brass cylinder, which is movable, and slides up and down the burner. Illuminating gas, however, is most convenient for blow-pipe experiments, and the Bunsen burners best suited for the purpose. (Fig. 2.) The burner rests on a foot, *a b*, into which a block, *c d*, is screwed. To this is attached the tube, *e f*, in which the gas coming through *k*, mixes with the air drawn through the holes in *c d*, and burns at *f* with an almost non-luminous flame.

For blow-pipe purposes, the tube *g h*, flattened at the top and slanted, is introduced, through which the gas passes without being mixed with the air, and burns with its usual luminous flame. The heating of substances in glass tubes and matrasses is best performed over a common spirit lamp, or Bunsen gas-burner.

§ 3. As supports, charcoal, platinum, and glass are principally used. Wood charcoal is in most cases the best support. It must be well burnt, and not scintillate or smoke; it must leave but little ash; charcoal of light wood, as alder and pine, has been found the best. Only those sides which show the rings of growth should be used.

Platinum is used whenever the reducing action of the charcoal acts injuriously. It is advantageously employed on all occasions where no reduction to the metallic state takes place, since the color of the flux is much better seen on the platinum than on charcoal. It is mostly used in the shape of wire, the end of which is bent so as to form a hook, which serves as support to the flux. (Fig. 3.) As foil, its use is very limited. A small platinum spoon or capsule, of from about 12 to 15 m. m. in diameter, is very convenient for fusing substances with bisulphate of potassa or nitre. (Fig. 4.)

Glass tubes, open at both ends, are used for calcination,

and for testing the presence of substances which are volatile at a high temperature. The tubes should be from 4 to 6 inches long. Of glass tubes, sealed at one end, or little matrasses, an assortment should always be kept on hand, since they are of very frequent use. (See Figs. 14, 15, 16, 17.)

§ 4. Of other apparatus, the most necessary are:

A mortar of agate or chalcedony, from one and a half to two inches in width, with pestle of the same material. (Fig. 5.)

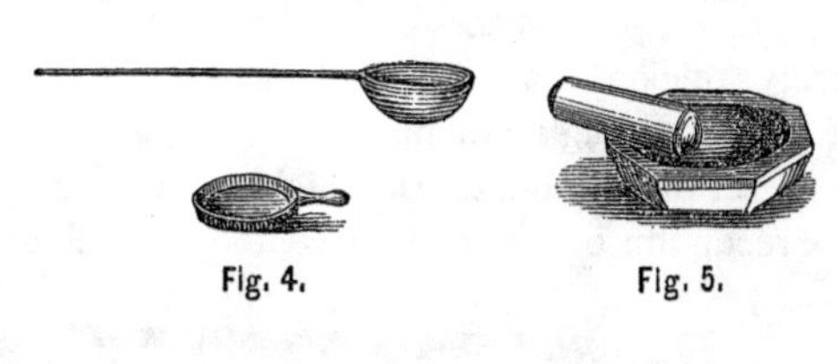

Fig. 4. Fig. 5.

Fig. 3.

Forceps of brass or German silver, with platinum points.

Forceps of steel.

A small hammer and anvil, both of steel and well polished.

A three-cornered file for cutting glass tubes, trying the hardness of minerals, etc.

A small magnet.

A pocket magnifying-glass.

Coal-borers. (Figs. 6, 7, 8.)

Cobalt glass and a prism with indigo solution. (Fig. 9.) This prism is made of plate glass, and filled with a solution prepared by dissolving 1 part of indigo in 8 parts of fuming sulphuric acid, adding 1500 to 2000 parts of water and filtering. In practice the prism is held close to the eye and moved horizontally, so as to allow the light of the colored flame to reach the eye through successively thicker portions of the absorbing medium.

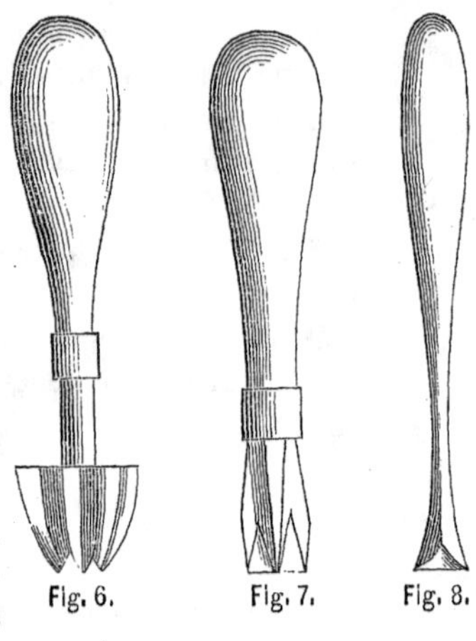

Fig. 6. Fig. 7. Fig. 8.

A set of watch-glasses, which are very convenient for the reception of the assay-piece, the metallic globules, etc.

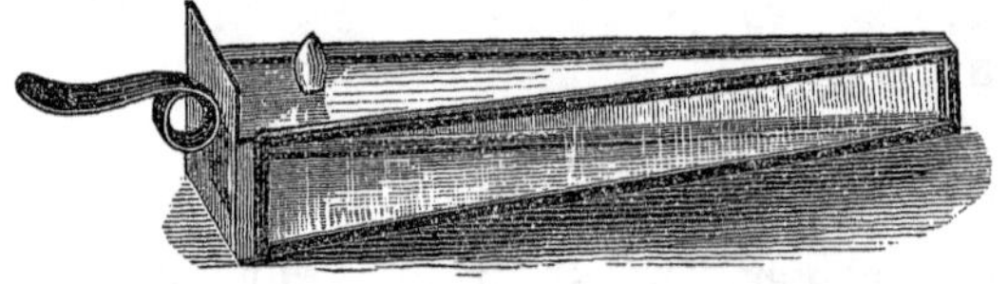

Fig. 9.

§ 5. Of reagents, Carbonate of Soda, Borax, and Salt of Phosphorus, are the most important ones; but there are others, which, though not so extensively used, still are indispensable for the detection of certain substances; others, the use of which is very limited, are omitted in this list. All should be as pure as possible.

Carbonate of Soda: The carbonate or the bicarbonate may be employed; it must be perfectly free from sulphuric acid, for the presence of which it may be tested, as shown § 107.

The neutral oxalate of potassa and the commercial [fused] cyanide of potassium deserve in many cases the preference, their reducing powers being superior to that of carbonate of soda.

Borax: The commercial article is purified by recrystallization, the crystals washed with distilled water, dried, and reduced to a coarse powder.

Salt of Phosphorus [phosphate of soda and ammonia]: 100 parts of crystallized phosphate of soda and sixteen parts of chloride of ammonium are dissolved in thirty-two parts of water; the solution is aided by heat, the liquid filtered while hot, and the crystals, which form on cooling, dried between blotting-paper. When pure it gives a glass which, on cooling, remains transparent; if this is not the case it must be purified by recrystallization.

Nitrate of Soda: Serves only as an oxidizing agent.

Bisulphate of Potassa: It is employed in the fused [anhydrous] state as a coarse powder; it must be kept in a bottle provided with a ground-glass stopper.

Vitrified Boracic Acid: It is employed in the state of a coarse powder.

Fluor-spar: Must be deprived of water by ignition; must be perfectly free from boracic acid, which may be tested as described § 61. It is convenient to keep in a separate bottle a mixture of one part of finely-powdered fluor-spar with four and a half parts of bisulphate of potassa.

Nitrate of Cobalt, in solution: It must be pure, free from alkali, sesquioxide of iron, and nickel. The solution should not be too concentrated, and as only one or two drops are used at a time, it is convenient to keep it in a bottle provided with a long stopper or pipette, for the purpose of dropping. (Fig. 10.)

Nitrate or Oxalate of Nickel: It must be perfectly free from iron and cobalt; it is tested with borax, with which it ought to produce a pure brown glass.

Oxide of Copper: It is best prepared by igniting the dried nitrate in a porcelain dish.

Fig. 10.

Chloride of Silver: It is prepared by precipitating a solution of nitrate of silver with hydrochloric acid, washing the precipitate, and making it into a thick paste with water, which is kept in a small glass-stoppered bottle. This reagent should not be used with platinum wire, since the silver fuses with the platinum to an alloy; thin iron wire is in this case substituted for the platinum. For each experiment a fresh hook should be made.

Pure Metallic Lead: It is easily obtained pure by decomposing a solution of the acetate by metallic zinc; the precipitate is repeatedly washed with water and then dried between blotting-paper.

Metallic Iron: In the shape of fine [pianoforte] wire.

Metallic Tin: Usually in the shape of foil, which is cut into strips and rolled up tightly.

Bone-ash: In the state of very fine powder for cupellation.

Test Paper: Blue and red Litmus Paper, and Brazil Wood Paper.

§ 6. If the analytical research is strictly confined to blow-pipe operations, the above enumerated reagents are sufficient; but if, as is sometimes advantageously done, some simple operations of the humid method of chemical analysis are called to aid, the list must be somewhat extended. The most important of these reagents, all of which must be kept in bottles with ground glass stoppers, are: Sulphuric Acid, Hydrochloric Acid, Nitric Acid,

Oxalic Acid, Hydrate of Potassa, Ammonia, Carbonate of Ammonia, Chloride of Ammonium, Molybdate of Ammonia, Ferrocyanide of Potassium, Ferridcyanide of Potassium, Bichloride of Platinum, Acetate of Lead, Sulphuretted Hydrogen Water, Sulphide of Ammonium, Alcohol, Distilled Water.

The principal auxiliary apparatus are: Test-tubes and Test-tube Rack, small Porcelain Dishes, small Beaker Glasses, Glass Funnels, Filter Stand, Filter Paper, Platinum Crucible, Glass Rods, Glass Plates for covering beakers, and either alcohol lamps or gas-burners.

CHAPTER II.

Of the Flame, and General Routine of Blow-Pipe Analysis.

THE flame of a candle consists of three distinct parts—the dark central zone or supply of unburnt gas surrounding the wick; the luminous zone or area of incomplete combustion; and the non-luminous zone or area of complete combustion. (Fig. 11.) In this outer zone the supply of oxygen is greatest, all the carbon is at once burnt, and the flame becomes non-luminous. The effect of producing a complete combustion at once throughout the flame is seen in the Bunsen gas-lamp, or burner. In this lamp (see Fig. 2) the gas issues from a small central burner, and passing up the tube draws air with it through the holes at the bottom of the tube; the mixture of air and gas can be lighted at the top of the tube, where it burns with a non-luminous flame. If the holes be closed, the gas alone burns with the ordinary bright flame. The blow-pipe flame may also be divided into two distinct parts—the *oxidizing* flame, where there is excess of oxygen, and the *reducing* flame, where there is excess of carbon; and these are distinguished by the same properties as the outer and inner zone of the candle flame.

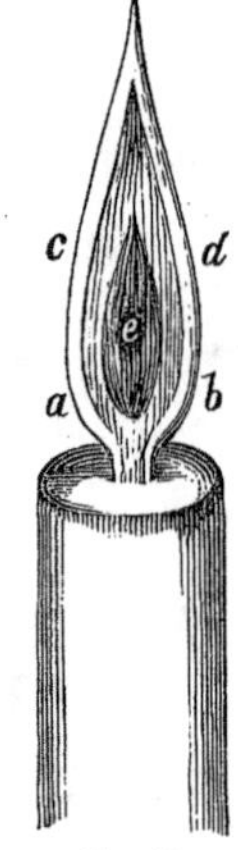

Fig. 11.

To produce the *oxidizing* flame (Fig. 12), the stream of gas should not be too strong. The jet of the blow-pipe is placed just within the flame near the slit in the tube, so that a strong current of air is thoroughly mixed with the gas, which forms an inner long blue flame, *a b*. The hottest part of the flame is just before the apex of this blue cone, *d*, where the combustion of the gases is most complete. For fusion, substances are exposed to this part, but for oxidation are placed a little beyond the apex.

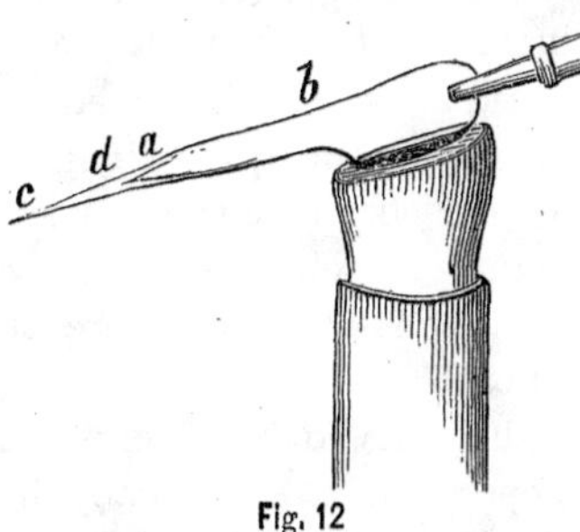

Fig. 12

The *reducing* flame (Fig. 13) is produced by so placing the blow-pipe that the jet just reaches the flame a little above the slit, and the current of air made to pass a little higher above the tube than in Fig. 12. The whole flame now appears as a long, luminous cone, *a b*, surrounded with a pale blue mantle, which extends to *c*. The most active part of the flame lies between *a* and *d*, somewhat nearer *a*. Any reducible metallic oxide placed at this point will be deoxidized or reduced, on account of the tendency of the free carbon in the flame to combine with the oxygen.

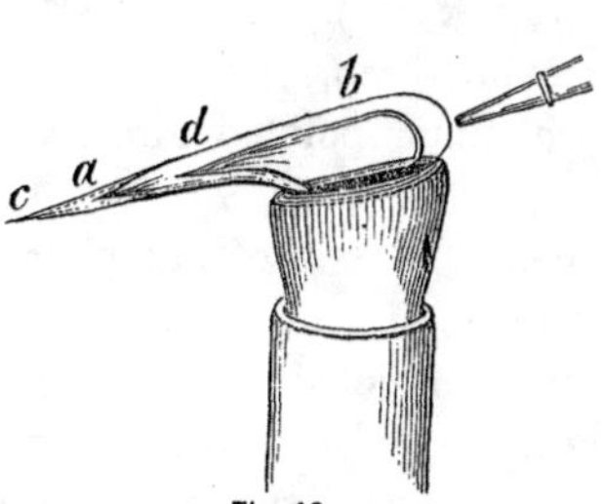

Fig. 13.

The current of air should be produced with the cheek muscles alone, and not with the lungs.

The trumpet-shaped mouth-piece is pressed against the lips, and breathing is effected through the nostrils. In this way a constant and regular blast may be produced, and, after a little practice, without any perceptible exertion or weariness.

§ 7. On examining a substance before the blow-pipe, with a view to determine its nature, or to ascertain the presence or absence of certain matter, it is advisable to follow a systematic course, composed of a series of operations, and to attentively observe the changes which the body undergoes under the influence of the various agents which are brought to act upon it. The various operations to which the assay is submitted are so many questions, to which the phenomena we observe constitute so many answers; and from their appearance, or non-appearance, we are able to draw definite conclusions as to the nature of the substance under examination.

The following order, and the rules to be observed in the execution of the various operations, are essentially the same as first pointed out and laid down by Berzelius.

1. Examination in a glass tube, sealed at one end, or a matrass, which is a glass tube sealed at one end and blown into a bulb. (Figs. 14 and 15.)

2. Examination in a straight or slightly bent glass tube open at both ends. (Figs. 16 and 17.)

3. Examination on charcoal by itself.

4. Examination in the platinum-pointed forceps, or on platinum wire by itself.

5. Examination with borax, and salt of phosphorus.

6. Examination with carbonate of soda.

7. Examination with solution of cobalt.

Regarding the size of the assay, a piece the size of a mustard seed will generally be found sufficient; larger pieces, without showing the reaction more distinctly, requiring so much more labor. In some cases, however, it is advantageous to employ a greater quantity, ex. gr. in

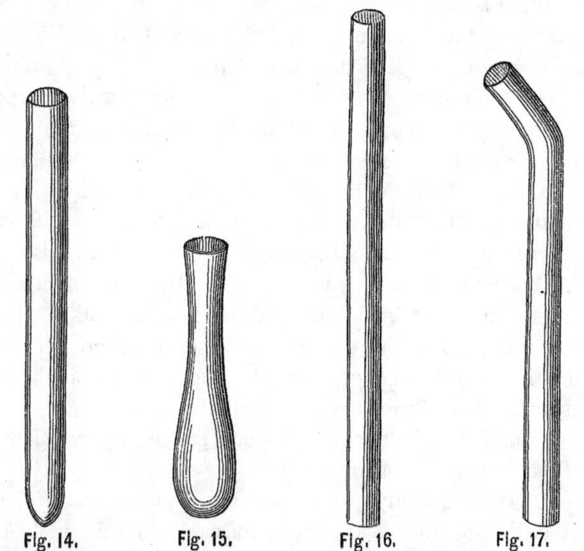

Fig. 14. Fig. 15. Fig. 16. Fig. 17.

reductions, or in heating in a glass tube; for the larger the metallic globule, and the greater the amount of the sublimate produced, the more readily can its nature be ascertained.

It is a good plan to place the lamp on a large piece of white paper, so that if a globule or portion of the specimen is dropped, it may be easily found.

Examination in a Closed Glass Tube, or a Matrass.

§ 8. The assay-piece is introduced into a small glass tube, sealed at one end, or into a small matrass, and heat applied by means of a gas or spirit lamp. The heat must at first be very low, but may be gradually raised to redness, if necessary. By this treatment we learn:

§ 9. 1. Whether the substance is entirely or partly volatile or not.

Among the phenomena to be observed, the following are deserving of particular attention:

The substance gives out water, which partly escapes and partly condenses in the colder portion of the tube. This points to the presence of a salt containing water of crystallization* [No. 22], or to the presence of a hydrate, or to such salts which contain water mechanically inclosed between the laminæ of the crystals [No. 18]; in this case the body usually decrepitates. The drops of condensed water are to be examined with test-paper: an alkaline reaction denotes the presence of ammonia, and acid reaction the presence of some volatile acid, as sulphuric, nitric, hydrochloric, hydrofluoric acid, etc.

§ 10. The substance gives out a gas or vapor. Those of most usual occurrence are:

a. **Oxygen,** easily recognized by placing a small piece of coal upon the assay, which burns brilliantly on being heated; points to the presence of a peroxide, nitrate, chlorate, bromate or iodate [No. 20].

b. **Sulphurous Acid,** easily recognized by its peculiar odor and action on blue litmus paper; indicates the presence of a sulphate or sulphite [No. 22].

c. **Sulphuretted Hydrogen,** recognized by its peculiar odor; indicates the presence of sulphides containing water.

d. **Nitrous Acid,** or peroxide of nitrogen, recognized by its deep orange-red color and acid reaction; indicates the presence of a nitrite or nitrate [No. 23].

* The numbers refer to list of substances following the introduction.

e. **Carbonic Acid**, recognized by causing a turbidity in a drop of lime-water suspended from a watch-crystal and exposed to the escaping gas; points to the presence of a carbonate.

f. **Cyanogen**, recognized by its peculiar odor and by burning with a crimson flame; indicates the presence of a cyanogen compound.

g. **Ammonia**, recognized by its odor and alkaline reaction; indicates the presence of an ammoniacal salt or of an organic nitrogenous substance; in the latter case, the mass usually blackens, and evolves at the same time either cyanogen or empyreumatic oils of offensive odor [No. 3].

h. **Hydrofluoric Acid** changes the color of test-paper, and also attacks the glass just above the assay, and makes it dull.

i. **Iodine** is indicated by violet fumes and its peculiar odor.

§ 11. The substance yields a sublimate. The sublimate is either white or possessed of a peculiar color. White sublimates are formed by

a. Many **Salts of Ammonia.**—On removing the sublimate from the tube, placing it on a watch-crystal, adding a drop of hydrate of potassa, and applying heat, ammonia is evolved [No. 19].

b. The **Chlorides of Mercury.** — The subchloride sublimes without previous fusion; the protochloride fuses first, then sublimes; the sublimate is yellow while hot, but becomes white on cooling [Nos. 39 and 40]. Oxide of mercury forms globules of metallic mercury.

c. **Oxide of Antimony.**—It fuses first to a yellow liquid, then sublimes; the sublimate consists of lustrous needle-shaped crystals [No. 24].

d. **Arsenous Acid.** — The sublimate consists of octahedral crystals [No. 25].

e. **Tellurous Acid.**— Shows a reaction similar to that of oxide of antimony, but requires a much higher temperature; the sublimate is amorphous.

f. **Osmic Acid** — Forms a sublimate of white drops, with a pungent, disagreeable odor.

Sublimates possessed of metallic lustre, so-called metallic mirrors, are formed by:

a. Metallic **Arsenic** and **Arsenides** containing more than one equivalent of arsenic to two of metal; also, some sulph-arsenides [No. 77]; cutting the tube below the sublimate, and exposing the mirror to gentle heat in the gas flame, the peculiar odor of arsenic is perceived.

b. **Mercury Amalgams,** and some salts of mercury; the sublimate consists of minute globules of metallic mercury, which, by friction with a piece of copper wire, readily unite to larger globules [No. 47].

c. Some alloys of **Cadmium.**

d. **Tellurium** — Only at a very high temperature; the sublimate consists of small globules, which solidify on cooling.

Sublimates possessed of distinct color are formed by:

a. **Sulphur** and **Sulphides** containing a large amount of sulphur; the sublimate is from deep-yellow to brownish-red while hot, but pure sulphur-yellow when cold [No. 75].

b. The **Sulphides of Antimony,** alone or in combination with other sulphides; the sublimate forms only at a very high temperature, and is deposited at a short distance from the assay-piece; it is black while hot, reddish-brown when cold [No. 74].

c. The **Sulphides of Arsenic** and some compounds of

metallic sulphides with arsenides; the sublimate is dark brownish-red while hot, but from reddish-yellow to red when cold [No. 80].

d. **Cinnabar.** — The sublimate is black, without lustre, and yields a red powder on being rubbed or scratched with a knife [No. 81].

e. **Selenium** and some selenides; the sublimate appears only at a high temperature, is of a reddish or black color, and yields a dark-red powder; at the open end of the tube the peculiar odor of selenium (resembling rotten horse-radish) is perceived [No. 87].

§ 12. 2. Whether the substance undergoes any change, or remains unaltered.

Many substances, under this treatment, suffer physical changes without being affected in their chemical constitution. A great many minerals, when heated, decrepitate; others phosphoresce, as fluorite and apatite. The most important of these physical changes is that of color: from white to yellow, and white again on cooling, points to oxide of zinc [No. 36]; from white to yellowish-brown, dirty pale-yellow on cooling, points to oxide of tin [No. 33]; from white to brownish-red, yellow when cold, and fusible at a red heat, points to oxide of lead [No. 72]; from white to orange-yellow or reddish-brown, pale-yellow when cold, and fusible at a bright red heat, points to teroxide of bismuth; from red to black, and red again on cooling, points to sesquioxide of iron [No. 64].

Examination in a Glass Tube Open at Both Ends.

§ 13. A fragment of the substance, sometimes in form of a powder, is introduced into the tube to a depth of about half an inch, the end to which it lies nearest slightly inclined, and heat applied. The air contained in the tube becomes heated; it rises, escapes from the upper end, and fresh air enters from below. In this manner a calcination is effected, and many substances which remained unchanged when heated in a matrass, yield sublimates or gaseous products when subjected to this treatment, owing to the formation of volatile oxides.

By this means we can easily detect the presence of the following substances:

§ 14. **Sulphur.**—Sulphurous acid is formed, which is characterized by its peculiar odor and action on moistened litmus paper [No. 75].

§ 15. **Arsenic.**—If present in sufficient quantity it yields a white and very volatile sublimate of arsenous oxide, consisting of minute octahedral crystals; by application of gentle heat it may be driven from one place to another [No. 77].

§ 16. **Antimony.**—White fumes of antimonous oxide are given out, which partly escape and partly condense in the upper part of the tube. The sublimate is a white powder, and may, if consisting of pure oxide of antimony, be volatilized by heat. In most cases, however, the oxidation proceeds farther, and antimonic oxide and antimonous oxide, a non-volatile white powder, is formed [No. 41].

§ 17. **Metallic Bismuth.**—It is converted into oxide, which condenses at a short distance from the assay, and which, by heat, may be fused to brownish globules [No. 43].

Mercury and Amalgams—Yield sublimates of metallic mercury, consisting of small globules [No. 47].

§ 18. **Tellurium and Tellurides.**—Tellurous acid is produced, which condenses in the upper part of the tube to a white non-volatile powder; on application of heat it fuses to colorless globules, thus distinguishing it from antimony.

Selenium and Selenides — Evolve a gaseous oxide of a peculiar odor, resembling that of rotten horse-radish [No. 87]; a sublimate of selenium gray near the assay and red at a distance is sometimes formed.

Examination on Charcoal alone.

§ 19. In examinations of this kind, particular attention must be paid to the odor of the escaping gases, and to the color and other properties of the rings, or coatings, which form on the charcoal around the assay-piece. The interior [reduction flame, R. Fl.] and exterior [oxidation flame, O. Fl.] cones of the flame acting in an opposite sense, the phenomena produced will be very different; hence two assays should always be made, exposing the substance first to the action of the O. Fl. and then to the action of the R. Fl. The following bodies undergo, when submitted to this treatment, characteristic changes.

§ **20. Arsenic.**—It is volatilized without previous fusion, the Ch. is covered with a white Ct., which is far distant from the assay-piece, and which is produced by both the O. Fl. and R. Fl.; the Ct. is very volatile, and is easily driven away by the Blp. flame, to which it imparts a light-blue color, emitting the peculiar alliaceous odor characteristic of arsenic [No. 42].

§ **21. Antimony.**—It enters readily into fusion and covers the Ch. with white oxide; the ring is not so far distant from the assay-piece as in the case of arsenic; it may be driven about by the O. Fl. and made to disappear with the R. Fl., which it colors a very pale-green, but is not so volatile as that of arsenic, and does not emit an alliaceous odor. Metallic antimony, when fused on Ch. and heated to redness, remains a considerable time in a state of ignition without the aid of the Blp., disengaging, at the same time, a thick white smoke, which is partly deposited on the Ch. around the metallic globule in white crystals of a pearly lustre [No. 41].

§ **22. Bismuth.**—It fuses readily in both flames and covers the Ch. with oxide, which is dark orange-yellow while hot and lemon-yellow when cold. The yellow Ct. is usually surrounded by a yellowish-white ring, consisting of carbonate of bismuth. The Ct. is somewhat nearer the assay than that of antimony; it may be driven away by both flames; but unlike antimony and lead, does not impart any color to the R. Fl. during the operation [No. 43].

§ **23. Lead.**—It fuses easily, and coats the Ch. in both flames with oxide, which is dark lemon-yellow while hot and sulphur-yellow when cold; in thin layers it is bluish-white and consists of carbonate. The Ct. is found at the same distance from the assay as that of bismuth; it may be driven away by both flames; when played upon with the R. Fl. it imparts to it an azure-blue color [No. 11].

§ **24. Cadmium.**—It fuses readily, and exposed to the O. Fl. it burns with a dark-yellow flame, emitting brown fumes of oxide, which cover the Ch. around and near the assay. This Ct. is very characteristic; it is, when cold, of a reddish-brown color, in thin layers orange-yellow; it is easily volatilized by both flames, without imparting a color to them. Beyond the coating a variegated border is sometimes seen [No. 44].

Indium.—It fuses readily, forming a coating very near the assay which is dark-yellow while hot, and yellowish-white when cold. It may be driven off with difficulty by the R. Fl., to which it gives a clear violet tint.

§ **25. Zinc.**—It fuses readily; exposed to the O. Fl. it burns with an intensely luminous greenish-white flame, emitting at the same time a thick white smoke, which, partly condensing on the Ch., rather near the assay, covers it with oxide, yellow while hot and white when cold. The Ct. when played upon with the O. Fl. becomes luminous, but does not disappear [No. 46].

§ **26. Tin.**—It fuses readily; exposed to the O. Fl. it is converted into oxide, which may be blown away and thus be made to appear as a Ct.; it is always found closely surrounding the assay-piece, is slightly yellow and luminous while hot, white when cold, and non-volatile in both flames. Exposed to the R. Fl. the molten metal retains its bright metallic aspect [No. 13].

§ **27. Silver.**—When exposed for a long time to the action of the R. Fl. it yields a slight dark-red Ct. of oxide [No. 45].

§ **28. Selenium.**—It fuses very readily in both flames with disengagement of brown fumes; at a short distance from the assay a steel-gray Ct. of a feeble metallic lustre is deposited; played upon with the R. Fl. it disappears with emission of a strong odor of rotten horse-radish, at the same time imparting to the flame a fine blue color [No. 87].

§ **29. Tellurium.**—It fuses very readily and coats the Ch. in both flames with tellurous acid; the Ct. is not very far distant from the assay; it is of a white color with a red or dark-yellow edge; played upon with the R. Fl. it disappears, imparting to the flame a green tinge.

§ 30. Besides the above-named metals there are some other substances which, when treated before the Blp. upon

Ch. cover it with coatings, which may be driven away when played upon with the O. Fl., and which show in many cases a great resemblance to the substances already noticed. Among the bodies possessing this property the following are the most frequently occurring ones:

The sulphides of potassium, sodium, lithium, and probably cæsium and rubidium.

The chlorides of ammonium, potassium, sodium, and lithium.

The chlorides of mercury, antimony, zinc, cadmium, lead, bismuth, tin, and copper.

The sulphides, bromides, and iodides of potassium and sodium.

Examination in the Platinum-pointed Pincers.

§ 31. This experiment serves a double purpose. It acquaints us with the degree of fusibility of the assay, and shows the presence or absence of such substances which possess the property of imparting to the flame a peculiar color. Many metals, the sulphides, and some other compounds act upon metallic platinum at a high temperature; the fusibility, etc., of such substances ought to be tested on Ch. Others, again, fuse so easily that they cannot be held a sufficiently long time between the pincers to observe the color which they impart to the flame; they are most conveniently attached to the hook of the platinum wire, which is best done by heating the wire to redness and then touching the powder of the assay with it; a sufficient quantity generally remains adhering to the wire.

Some minerals decrepitate violently as soon as they are touched with the flame; in such cases, Berzelius advises to powder the substance very finely in an agate mortar with addition of a little water, to place one or two drops of the mixture on a piece of Ch., and to gently heat it by means of the Blp. flame until the mass lies loosely upon the Ch.; it may then be taken up and held by the pincers. The same process is advantageously employed with substances which fuse only at a very high temperature. In all other cases the substance is roughly powdered and a thin piece which shows prominent edges selected for the experiment.

The assay is exposed to the action of the inner cone of the flame, when the outer cone may exhibit the following changes of color: yellow, violet, red, green or blue.

§ 32. **Yellow.** — Soda and its salts cause an enlargement of the outer flame, and impart, at the same time, an intense reddish-yellow color [No. 18]. The presence of other substances which also possess the property of coloring the flame, but not in so high a degree, does not prevent the reaction. Silicates containing soda exhibit the same phenomenon to a smaller or greater extent, according to their degree of fusibility and the amount of soda which they contain [No. 62]. With many salts of soda, which do not exhibit the reaction very distinctly, it can be produced by mixing the salt with some chloride of silver to a paste (v. § 5), fastening it to the hook of a thin iron wire, and then exposing it to the action of the inner flame.

§ 33. **Violet.** — Potassa and most of its salts, with the exception of borate and phosphate, impart to the outer flame a distinct violet color [No. 15]. Also the salts of rubidium and cæsium and the compounds of indium, but these are rare as compared with the potassa. The presence of a salt of soda prevents the potassa reaction, in which case the flame is viewed through blue cobalt glass, or a solution of indigo. Lithium also destroys the potassa flame, unless in very minute quantities. Silicates of potassa must be free from soda and lithia, and easily fusible, at least on the edges.

§ **34. Red.**—Lithia and its salts impart to the outer flame a fine carmine-red color [No. 63]; the chloride of lithium shows the reaction better than any other salt. The presence of a salt of potassa does not prevent the reaction; the presence of even a small quantity of a salt of soda changes the color to yellowish-red, and a larger quantity prevents the reaction entirely.

Chloride of strontium and some other salts of strontia, ex. gr. the carbonate and the sulphate, color the outer flame, immediately or after a while, carmine-red [No. 57]. The presence of considerable baryta prevents the reaction. The carbonate and sulphate of strontia show the reaction remarkably well when mixed with chloride of silver and heated on iron wire (v. § 5).

Chloride of calcium, calcareous spar, many compact limestones, and fluorite, color the outer flame, immediately or after a while, red; the color is not so intense as that produced by strontia. Gypsum and anhydrite impart at first a pale yellow, afterward a red color of little intensity [No. 55]. Fluorite gives at first a yellowish flame, but afterward an intense yellowish-red [No. 6].

§ **35. Green.**—Baryta. Chloride of barium, carbonate and sulphate of baryta, color the outer flame yellowish-green. The presence of lime does not prevent the reaction [No. 58].

Oxide of copper and some of its salts, ex. gr. the carbonate and nitrate, impart to the outer flame a fine emerald-green color. Compounds of iodine and copper, and some silicates containing copper, ex. gr. dioptase and chrysocolla, act in the same manner [No. 73]. Thallium, fused on charcoal, also its salts, give an intense green flame.

Phosphoric acid, phosphates, and minerals containing phosphoric acid, especially if moistened with sulphuric acid, impart to the outer flame a bluish-green color [No. 3].

Boracic acid colors the outer flame yellowish-green (greenfinch color) [No. 5]; if a small quantity of soda is present the color is mixed with yellow. Minerals containing boracic acid should be pulverized and moistened with sulphuric acid.

Molybdic acid, oxide of molybdenum, and molybdenite color the outer flame yellowish-green, in which the yellow is stronger than in baryta [No. 83].

Tellurous acid fuses, emits white fumes, and colors the outer flame green.

§ 36. **Blue.** — Arsenic and some arsenides, ex. gr. smaltite and niccolite [No. 82], when heated on Ch., impart a light-blue color to the outer flame. Some arsenates, ex. gr. scorodite and erythrite, exhibit the same phenomenon in the forceps.

Antimony, fused on Ch. in R. Fl., is surrounded by a feeble greenish-blue flame [No. 41].

Metallic lead, fused on Ch. in R. Fl., is surrounded by an azure-blue flame. Many salts of lead, heated in the forceps or on platinum wire, impart an intense azure-blue color to the outer flame [No. 11].

Chloride of copper colors the outer flame intensely azure-blue; after a while the color becomes green, owing to the formation of oxide of copper [No. 37].

Bromide of copper colors the outer flame greenish-blue; after a while the color changes to green.

Selenium, fused on Ch. in R. Fl., vaporizes with an azure-blue flame.

Examination with Borax and Salt of Phosphorus.

§ 37. The examination of the assay with borax and salt of phosphorus is eminently adapted to detect the presence of metallic oxides, a great number of them possessing the property of being at a high temperature dissolved by these fluxes with a peculiar color. Unoxidized metals and metallic sulphides, arsenides, etc., differ in this respect very materially from the pure oxides; hence it is necessary, before performing the experiment, to convert all such substances into oxides. This is effected by calcination, or roasting on Ch. or in an open glass tube. The finely powdered assay is placed on Ch. and alternately treated with the O. Fl. and R. Fl., and this process is repeated until the substance no longer emits, while in the incandescent state, the odor of sulphur or arsenic. The heat must never be raised so high as to cause fusion, and between every two succeeding calcinations the assay should be taken from the Ch. and freshly powdered.

The experiment with borax is generally made on platinum wire, where the color of the bead is more readily observed; Ch. is used only in such cases where the substance under examination contains metallic oxides which are easily reduced. It is not sufficient to observe the color of the bead after cooling; but all changes of color which take place during the action of the flame, and through all the various stages of cooling, should be carefully noticed.

Some substances possess the property of forming a limpid glass, with borax, which preserves its transparency on cooling, but which, if slightly heated in the O. Fl., becomes opaque when the flame strikes it in an unequal or intermittent manner. This operation has received the name of "flaming," and any substance thus acted upon is said to become "opaque by flaming."

The third and fourth columns of Tables I. and II. exhibit the behavior of the most important oxides to borax and salt of phosphorus.

In Table III. the oxides are arranged with reference to the color which they impart to the beads in O. Fl. and R. Fl. It will be noticed that the colors produced in the S. Ph. are different from those produced in borax.

Examination with Carbonate of Soda.

§ 38. In subjecting a body to the treatment of Sd. we have to direct our attention to two points.

Some substances unite with Sd. to fusible compounds, others form infusible compounds, and others again are not acted upon at all; in the last case the Sd. is absorbed by the Ch. and the assay is left unchanged. With Sd. form fusible compounds with effervescence:

§ **39. Silicic Acid.**—It fuses to a transparent glassy bead which, after cooling, remains transparent if the Sd. has not been added in excess [No. 54].

Titanic Acid.—It fuses to a transparent glassy bead which, when cold, is opaque and of crystalline structure [No. 65].

Tungstic and Molybdic Acids.—The mass, after the union has been effected, is absorbed by the Ch. [No. 31 and No. 34].

Tantalic, vanadic, and the acids of niobium, also yield fusible compounds.

The salts of baryta and strontia form with Sd. fusible compounds which are absorbed by the Ch. [No. 57 and No. 58].

§ 40. The second point to be observed is the elimination of metallic matter. Of the metallic oxides, when treated with Sd. on Ch. in R. Fl., are reduced: the oxides of the noble metals and the oxides of arsenic, antimony, bismuth, indium, cadmium, copper, cobalt, iron, lead, mercury, nickel, tin, zinc, molybdenum, tungsten, and tellurium. Of these, arsenic and mercury vaporize so rapidly that frequently not even a coating is left on the Ch. Antimony, bismuth, cadmium, lead, zinc, and tellurium are partly volatilized and form distinct coatings on the Ch. The non-volatile reduced metals are found mixed up with the Sd. To separate them from the adhering Sd. and Ch. powder, we may proceed in the following manner:

The fused mass of Sd. and metal, and the portion of the Ch. immediately below and around the assay, is placed in the little agate mortar, rubbed to powder, the powder mixed with a little water, and stirred up. The heavy metallic particles settle to the bottom, part of the Sd. dissolves, and the Ch. powder remains suspended in the water. The liquid is carefully poured off, and the residue treated repeatedly in the same manner until all foreign matter is removed. The metal remains behind as a dark heavy powder, or, when the metal is ductile and easily fusible, in the shape of small flattened scales of metallic lustre. These may be examined with the magnifying-glass, and also with the magnet. If the substance under examination contains several metallic oxides, the metallic mass obtained is usually an alloy, in which the several metals may be recognized by processes to be described hereafter. It is only in some exceptional cases that separate metallic globules are obtained, ex. gr. in substances containing iron and copper.

For a more detailed account of the behavior of the

various metallic oxides under this treatment, see the second column of Tables I. and II.

§ 41. The examination with Sd. is usually performed on Ch. in the R. Fl., and, as a general rule, the flux is added successively in small portions. It is sometimes better to form the pulverized assay into a paste with moistened soda before placing it upon the coal. This is particularly necessary when the assay is to be tested for its fusibility with Sd., since a great many minerals, etc., behave very differently with different quantities of the flux.

§ 42. Instead of carbonate of soda, the *neutral oxalate of potassa* or *cyanide of potassium* may be advantageously used for all experiments of reduction, since these reagents exercise a more powerful reducing action than common Sd. They are, for this reason, frequently employed when the presence of such metallic oxides is suspected, whose conversion into metals require high temperatures and the aid of a very efficient deoxidizing agent.

A list of the oxidized minerals arranged according to their fusibility and behavior with carbonate of soda may be found under its appropriate head.

Examination with Solution of Cobalt.

§ 43. A few substances, when moistened with a solution of nitrate of cobalt and exposed to the action of the O. Fl., assume a peculiar color. The use of this test is, however, very limited, since the reaction can only clearly be seen in such bodies which, after having been acted upon by the O. Fl., present a white appearance, or nearly so.

§ 44. Substances which are sufficiently porous to absorb a liquid, are merely moistened with a drop of S. Co., placed into the platinum-pointed forceps, and treated with the O. Fl. Other substances must be powdered, the powder placed on Ch., moistened with a drop of S. Co., and treated as above. The color can only be distinguished after cooling. A bluish color, of more or less purity, but rather dull, indicates the presence of alumina [No. 21]; and a pale-reddish color [flesh-color] that of magnesia [No. 59]. It must, however, be borne in mind, that the alkaline and some other silicates, when heated with S. Co. to a temperature above their fusing point, also assume a blue color, owing to the formation of silicate of cobalt. In testing for alumina, therefore, the heat must not be raised so high as to cause fusion of the assay. In testing for magnesia this precaution is not necessary; on the contrary, the color will appear the brighter and the more distinct, the higher the temperature to which the assay was exposed. The alumina and magnesia reactions are prevented by the presence of colored metallic oxides, which generally produce a gray or black mass, unless present in too minute quantity.

§ 45. Among the oxides of the heavy metals, those of zinc and tin assume characteristic colors with S. Co. The reaction is best seen when the assay, alone or mixed with Sd., is exposed to the R. Fl. on Ch. The ring of oxide which is deposited around the assay is then moistened with S. Co. and treated with the O. Fl. Oxide of zinc takes a fine yellowish-green, and oxide of tin a bluish-green color [No. 36 and No. 33].

§ 46. Besides the compounds above mentioned there are some others which, when exposed to the action of S. Co. and Fl., experience a change of color. These bodies are either of very rare occurrence, or the change produced in them is not sufficient to be of much importance. In fact only a few colorations are of much use in the determination of substances, those of alumina, magnesia, zinc, and tin.

The following list of compounds gives the more definite colorations:

a. Blue, alumina, silica.

b. Violet, zirconia (dirty-violet), phosphate and arsenate of magnesia (fuse at the same time).

c. Flesh color, magnesia, tantalic acid.

d. Brownish-red, baryta.

e. Gray, strontia, lime, glucina, niobic acid.

f. Green, oxide of zinc (yellowish-green), of tin (bluish-green), titantic acid (yellowish-green), hyponiobic acid (dirty green), antimonic acid (dirty dark-green).

CHAPTER III.

Special Reactions for the Detection of Certain Substances when in Combination with Others.

§ 47. THE preceding chapter and accompanying table show the changes which many of the simple chemical compounds undergo when heated, or when treated with the usual blow-pipe reagents. The reactions are sufficiently characteristic to distinguish the various compounds from each other, so that, when any of the above-named substances in a pure state is under examination, there is no difficulty to determine its nature. This, however, is not of frequent occurrence, and in the majority of cases the body to be tested will be of a more complex nature. The results of the experiments will vary accordingly. For instance, an ore of cobalt, containing iron, will not impart to the bead of Bx. or S. Ph. in the O. Fl. a blue color, but a green one, resulting from the mixture of the blue of cobalt and the yellow of iron; lead, when accompanied by antimony, deposits a dark-yellow coating on Ch. resembling that of bismuth, etc. In such cases we may often, by attentively observing all the phenomena which present themselves, and by carefully comparing the results obtained by the various experiments, detect many, if not all, of the components of the substance under examination. Sometimes we attain this end quicker by varying the order, or by introducing auxiliary agents into the series of experiments; and in other cases, again, it is only to be arrived at by subjecting the assay to treatments different from those mentioned in the preceding pages.

This chapter contains the principal reactions for the detection of substances which require the application of peculiar agents, and the methods for ascertaining the presence of certain bodies when in combination with others. The alphabetical arrangement will be found of practical use.

§ **48. Ammonia.**—Small quantities of ammonia are best detected by mixing the powdered assay [No. 19] with some carbonate of soda or caustic potassa, introducing the mixture into a glass tube, sealed at one end, and applying heat. The escaping gas is characterized by its odor, and by its action on reddened litmus paper. White clouds are formed if a glass rod moistened with hydrochloric acid is held before the end of the open tube. From the appearance of this reaction we are, however, not authorized to infer the preëxistence of ammonia in the assay, since from organic matter containing nitrogen, when subjected to this treatment, ammonia is evolved as a product of decomposition.

Antimony.—The reactions of antimony and its compounds, see § 11, § 16, § 21, § 36, and Table II., 1.

§ 49. In presence of lead or bismuth, antimony cannot be detected by its Ct. alone on Ch. In this case the metallic compound [No. 48 or No. 85] is treated with vitrified boracic acid on Ch., the flame being so directed that the glass is always kept covered with the blue cone, the metallic globule being on the side; by this means the metals become oxidized, the oxides of lead and bismuth are absorbed by the boracic acid, and the oxide of antimony will form a ring on the Ch., provided the temperature was not raised too high.

§ 50. When combined with metals from which it is not easily separated, ex. gr. copper, the evaporation of the antimony takes place so slowly that no distinct Ct. is produced. In this case the assay [No. 86] is treated with S. Ph. on Ch. in the O. Fl., until the antimony, or at least part of it, has become oxidized and entered into the flux. The glass is now removed from the metallic globule and treated on another place of the Ch. with metallic tin in the R. Fl.; the presence of antimony will cause the glass to turn gray or black on cooling [Table II., 1]. Bismuth behaving under these circumstances in precisely the same manner, the presence of this metal makes the reaction not decisive for antimony. The humid way has then to be resorted to. See § 59.

§ 51. When the oxides of antimony are accompanied by such metallic oxides which, when reduced on Ch., fuse with the metallic antimony to an alloy, as is ex. gr. the case with the oxides of tin and copper, the latter cannot be recognized by a simple reduction. The oxides have to be treated with a mixture of Sd. and Bx. on Ch. in the R. Fl. The little metallic globules are separated from the flux, and fused with from three to five times their own volume of pure lead and some vitrified boracic acid in the R. Fl., care being taken to play with the flame only on the glass. Oxide of antimony is volatilized, depositing the characteristic ring, while the oxides of the other metals are absorbed by the boracic acid.

§ 52. The sulphides of antimony, when heated in the open glass tube, show the reaction mentioned § 16. When accompanied by sulphide of lead [No. 89], only a small part of the antimony is converted into oxide, which sublimes; the remainder is changed into a white powder consisting of a mixture of antimonic acid and oxide of antimony, sulphate of lead, and antimonate of lead. When a compound containing sulphide of lead or bismuth, besides sulphide of antimony, is heated on Ch. in the R. Fl., a Ct. is deposited consisting of oxide of antimony mixed with sulphate of lead or bismuth, and, nearer to the assay, a yellow one of the oxides of lead or bismuth; how in such a case the presence of antimony may be ascertained v. § 87.

§ 53. To detect a small amount of sulphide of antimony in sulphide of arsenic, Plattner strongly recommends the following method, by which he obtained very decisive and satisfactory results: The assay [No. 88] is introduced into a glass tube, sealed at one end, and gently heated; the sulphide of arsenic is volatilized, and the greater part of the sulphide of antimony remains as a black powder in the lower end of the tube; this end is cut off, and the black substance taken out and transferred to a tube open at both ends. By applying heat the characteristic antimony-reaction will appear.

Arsenic. — The reactions of arsenic and its compounds, see § 11, § 15, § 20, § 36, and Table II., 2.

§ 54. All metallic arsenides yield, when heated in the open glass tube, a sublimate of arsenous oxide (v. § 15), and most of them evolve a garlic odor (v. § 20), when heated on Ch. in R. Fl. [No. 77]. Some metals, ex. gr. nickel and cobalt, have a great affinity for arsenic, so that, when only a small quantity of the latter is present, the characteristic odor is not observable; in such cases it is sometimes produced when the metallic compound is fused on Ch. with some pure lead in the O. Fl.

§ 55. The sulphides of arsenic, heated in the open glass tube, evolve sulphurous acid and yield a sublimate of arsenous oxide. To show in a very decisive manner the presence of arsenic in any of its combinations with sulphur, the powdered assay [No. 80] is mixed with four volumes of neutral oxalate of potassa and a little charcoal powder, or with six parts of a mixture of equal parts of cyanide of potassium and carbonate of soda, the mass introduced into a tube sealed at one end, and heat applied, at first very gently but gradually raised to redness. A ring of metallic arsenic will be deposited in the colder part of the tube. (Fig. 18.)

Fig. 18.

§ 56. When sulph-arsenides are heated on Ch., the whole of the arsenic, especially when only small quantities are present, may pass off in combination with sulphur; but when such compounds [No. 88] are mixed with from three to four parts of neutral oxalate of potassa or cyanide of potassium and exposed to the R. Fl., sulphide of potassium is formed and the arsenic escapes with its peculiar odor, if not combined with cobalt or nickel.

§ 57. To detect a very small quantity of arsenous acid, the following way may be pursued: a glass tube provided with a small bulb at one end is close above it narrowly drawn out; the assay [No. 38] is introduced into the bulb, and a charcoal splinter placed into the tube; the narrow aperture through which the tube communicates with the bulb prevents the Ch. from coming in contact with the substance. The tube is then heated to redness at the place where the charcoal splinter lies, and as soon as this is incandescent, heat is also applied to the bulb. (Fig. 19.) The arsenous oxide is volatilized, and its vapors, while passing over the red-hot charcoal, become reduced and deposit a black metallic ring of arsenic in the colder part of the tube. By cutting the tube below the ring and heating this part by the flame of a gas-lamp, the arsenic is volatilized, thereby emitting its characteristic odor.

§ 58. To show the presence of arsenic in arsenites and arsenates, it will in most cases be sufficient to mix the

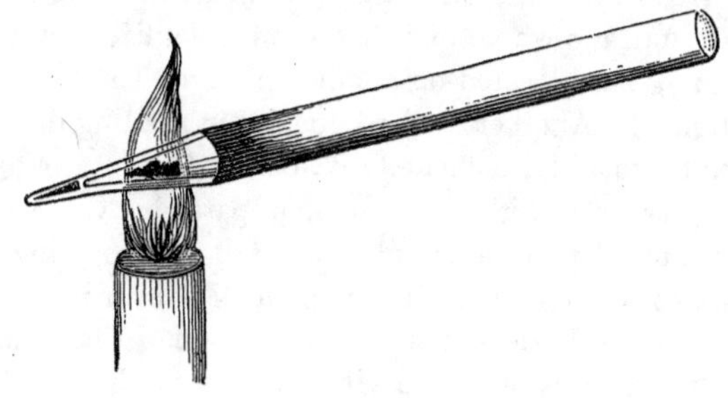

Fig. 19.

substance [No. 38] with carbonate of soda and heat it on Ch. in R. Fl. Sometimes it is necessary to treat the assay with a mixture of carbonate of soda and cyanide of potassium in the manner mentioned, § 55; and in other cases again, where but small quantities of arsenous or arsenic oxide are combined with metallic oxides which are readily reduced, recourse must be had to the humid way.

Bismuth.—The reactions of bismuth and its compounds, see § 12, § 17, § 22, and Table II., 3.

§ 59. Bismuth, when alloyed with other metals, or when as sulphide in combination with other sulphides, is in many cases, and most especially so when accompanied by lead or antimony, not to be detected with certainty by the ring which it deposits on Ch. In such a case the assay [No. 49] is treated on Ch. until a copious yellow Ct. is formed. The Ct. is carefully scraped off from the Ch. and dissolved in S. Ph. on platinum wire with the O. Fl. The colorless bead is removed from the wire, placed on Ch., a little metallic tin added, and the whole exposed to the R. Fl. If bismuth was present, the glass assumes, on cooling, a dark-gray or black color. The oxides of antimony showing the same behavior, the assay, if not quite free from antimony, has to be treated on Ch. in the O. Fl. until the whole of it has been volatilized, and the remaining mass treated on another piece of Ch. as above mentioned. Another method consists in roasting the substance thoroughly, but carefully, on coal, to prevent sintering, fusing it with three to four parts by volume of bisulphate of potassa on platinum foil, and treating the mass with water in a small porcelain dish over the lamp, until it is wholly detached from the spoon. The soluble sulphates are dissolved, leaving neutral sulphate of lead and basic sulphate of bismuth; only a small part of the bismuth is dissolved as neutral sulphate. Antimony, if present, also remains behind as acid.

The clear solution is decanted, the residue boiled with water, a few drops of sulphuric and nitric acid added, when the sulphate of bismuth dissolves, leaving a residue of sulphate of lead, with oxide of antimony if present. After filtration the bismuth is thrown down from the warm filtrate with S. Ph., which is collected and tested with S. Ph. The bead on platinum wire is colorless or only tinged with yellow, but on coal with tin in R. Fl. becomes dark-gray on cooling, like oxide of bismuth, and may be recognized as such on coal.

According to Von Kobell, any compound of bismuth treated before the Blp. with a mixture of equal parts of iodide of potassium and sulphur on a large coal, gives a beautiful and very characteristic red coating, at quite a distance from the assay.

Cornwall * suggests the following method to detect bis-

* American Chemist, March, 1872.

muth in presence of lead and antimony: To the mixture of the three oxides, an equal volume of sulphur is added, and the whole treated before the Blp. in a deep cavity on coal with the blue flame for a few moments. The resulting fused sulphides are removed to a flat coal and treated alternately with the O. Fl. and R. Fl. until the antimonial fumes have nearly ceased, and an impure blue lead flame appears. The residue is powdered, and an equal part of a mixture of 1 part of iodide of potassium and 5 of sulphur, by weight, added. This is then heated in an open tube about four inches long and not less than one-third of an inch wide, over a Bunsen gas-burner, or spirit lamp. A distinct red bismuth sublimate is formed, about one-third of an inch above the yellow sublimate of lead.

Care must be taken not to confound with the bismuth sublimate a sublimate of iodine, which may condense on the upper part of the tube, but at a greater distance from the assay.

§ **60. Boracic Acid.**—With many borates, which do not impart to the outer flame the peculiar yellowish-green color [v. § 35], this reaction may be produced by reducing the substance [No. 2] to powder, adding a drop of concentrated sulphuric acid, fastening the mixture into the hook of the platinum wire, and playing on it with the blue cone of the flame.

§ 61. Another way, and by which even a very small quantity of boracic acid in salts and minerals may be detected, is: to reduce the substance to a very fine powder, to mix it with from 3 to 4 parts of a mixture of 4½ parts of bisulphate of potassa and 1 part of fluor-spar, perfectly free from boracic acid, and to knead the whole with a little water into a thick paste. This mass is then fastened to a platinum wire, and exposed to the blue cone of the flame. While the mass enters into fusion fluoboracic acid is formed, which, on escaping, colors the flame intensely yellowish-green. The reaction appearing sometimes only for a few seconds, the flame should be very attentively watched during the whole time of the experiment. According to Merlet, it is often necessary, in order to obtain a sure result, to employ with one part of the assay three or four of the flux.

§ **62. Bromine.**—Bromides treated with S. Ph. and oxide of copper on platinum wire, or treated with sulphate of copper on silver foil, show the same reaction as chlorides (v. § 66), with this difference, that the blue color of the outer flame is rather greenish, especially on the edges [No. 16]. When the bromine is all driven off the green flame of the copper alone remains.

§ 63. To discriminate bromides from chlorides more distinctly, the bromide is fused with bisulphate of potassa, both in the anhydrous state, in a small matrass with long neck. Sulphurous acid is evolved, and the matrass is filled with yellow vapors of bromine, characterized by their peculiar odor. The color of the gas is only clearly seen by daylight. Bromide of silver may be distinguished from chloride of silver by the asparagus-green color which it assumes when exposed to the sunlight after fusion with bisulphate of potassa.

The presence of iodine, on account of its violet vapors, renders the bromine reaction somewhat uncertain.

Cadmium.—The reactions of cadmium and its compounds, see §§, 11, 24, and Table II., 4.

§ 64. To detect a very small quantity of cadmium, one per cent. or less, in zinc or its ores, the pulverized assay is mixed with Sd. and exposed for a short time to the R. Fl. on Ch. A distinct Ct. of oxide of cadmium is deposited. The zinc being less volatile, volatilizes only with continued blowing [No. 53].

§ **65. Chlorine.**—Some oxide of copper is dissolved by means of the O. Fl. in a bead of S. Ph. on platinum wire, until the glass is nearly opaque. Some grains of the pulverized assay [No. 18] are then made to adhere to the bead, and both heated with the tip of the blue cone of the flame. If chlorine is present the flame now assumes an intense azure-blue color, owing to the formation of chloride of copper (v. § 36). This test is very delicate, and will show the presence of a very minute quantity of chlorine. Bromine produces a similar flame.

§ 66. Another method is to place on silver-foil some protosulphate of iron, or some sulphate of copper, to moisten it with a drop of water, and then to add the assay [No. 18]. After a while the silver will be found blackened. Substances which are insoluble in water have previously to be fused with a little Sd. on platinum wire, to form a soluble chloride [No. 10].

Chlorides, when moistened with sulphuric acid and exposed to the Blp. flame, impart to it a faint green coloration, which, however, is generally confined to the inner cone, and is quantitatively of much less intensity than that produced with borates. A small amount of boracic acid, when occurring together with a chloride, cannot, therefore, be detected by the method mentioned § 60.

§ **67. Chromium.**—Oxide of chromium gives very characteristic reactions with the fluxes on platinum wire (v. Table II., 6), but when accompanied by a large quantity of iron, copper, or other substances which also intensely color the Bx. and S. Ph. beads, the chromium color frequently becomes very indistinct.

§ 68. In such a case, and when the chromium is not in combination with silicic acid, its presence may be detected in the following manner: The assay-piece [No. 71] is reduced to a fine powder and mixed with about twice its own volume of a mixture of equal parts of Sd. and nitre. The mass is fastened into the hook of a thick platinum wire, or placed into a small platinum spoon, and treated with a powerful O. Fl. An alkaline chromate is formed which is dissolved in water, the solution supersaturated with acetic acid, boiled, and a crystal of acetate of lead added. If chromium was present, a yellow precipitate of chromate of lead will appear. The precipitate may be collected on a filter and tested in the Bx. and S. Ph. beads, when the characteristic chromium-reactions will be produced.

Silicates which contain only a little chromium, but much iron or other coloring oxides of metals, are not decomposed by nitre. In this case a pulverized mineral is fused on coal in the O. Fl. with one to one and a half times its volume of soda, and one-half to three-fourths parts of borax to a clear bead; this is pulverized and evaporated to dryness with hydrochloric acid. The chlorides thus formed are dissolved in water, the silica filtered off, the protochloride of iron in solution changed to sesquichloride by boiling with a few drops of nitric acid, and the bases, sesquioxide of chromium, iron, alumina, etc., precipitated by ammonia from the acid solution. The precipitate is collected on a filter, washed, and fused with soda and nitre as above. By this means chromates of the alkalies are formed, which can be decomposed by acetic acid and acetate of lead as already described.

Cobalt.— The reactions of cobalt, see Table II., 7.

§ 69. To detect cobalt, when in combination with other metals, v. § 83.

To show its presence in arsenides, the assay [No. 78] is placed on Ch. and heated until fumes of arsenous acid no longer are emitted. (Lead and bismuth, if present, form the characteristic coatings.) Bx. is now added and the heat continued until the glass appears colored. If the color is not pure blue, the presence of iron is indicated. The glass is in this case removed from the globule, and the latter treated repeatedly with fresh quantities of Bx. until the pure cobalt-color is obtained. Nickel and copper, if present, do not enter into the flux before the whole of the cobalt is oxidized. If we wish to ascertain the presence of these metals, the glass which is colored by cobalt is removed from the globule, and the latter treated with fresh portions of Bx. in the O. Fl. until the color of the bead becomes brown, indicative of nickel. The glass is again removed and the globule treated with S. Ph. in the O. Fl.; when copper is present the bead assumes a green color, which remains unaltered on cooling. Treated with tin on Ch. the glass turns opaque and red from suboxide of copper.

§ 70. To detect cobalt in sulphides, the assay [No. 79] is heated on Ch. in the R. Fl. until all volatile substances are driven off, the remaining mass reduced to powder, well calcined, and the calcined mass treated with Bx. on Ch. in the O. Fl. If cobalt is the only coloring metal present, the bead will exhibit a pure blue color; a small addition of iron will make the glass appear green while hot, but blue when cold. Copper and nickel, when present to some extent, will prevent the cobalt-color being distinctly seen. The bead is in this case exposed to the R. Fl. until it appears transparent and flows quietly; the oxides of copper and nickel are by this means reduced, and the pure color of cobalt, or that of cobalt mixed with iron, becomes apparent. The separation of the metals may be promoted by adding a little pure lead, and the substance freed from an excess, by treating it alone on coal, after which it is fused in the O. Fl. with S. Ph. to detect nickel and copper.

Copper. — The reactions of copper and its compounds, see §§ 35, 36, and Table II., 8.

§ 71. The red color which copper imparts to the Bx. or S. Ph. bead, when heated on Ch. in the R. Fl. in contact with tin (v. Table II., 8), is very characteristic, and will in most cases clearly show the presence of this metal. But if only a small quantity of copper is associated with other metals, the reaction is not easily obtained; in this case we may proceed as follows:

The assay [No. 89, or No. 86, or No. 85] is placed on Ch. and played upon with the O. Fl. until antimony and other volatile metals are driven off. Some vitrified boracic acid is fused on Ch. to a glassy globule, the assay placed close to it, and the whole covered with a large R. Fl. When the metallic globule begins to assume a bright metallic surface, the flame is gradually converted into a sharply-pointed blue cone, which is made to act only on the glass, leaving the metallic globule untouched, and so situated that it touches the glass on one side, and on the other side is in close contact with the Ch. During this process lead, iron, cobalt, part of the nickel, and such of the more volatile metals as were not entirely removed by the previous calcination, as bismuth, antimony, zinc, etc., become oxidized, and their oxides partly volatilized and partly absorbed by the boracic acid. The remaining metallic globule is then removed from the flux and treated on Ch. with S. Ph. in the O. Fl., when the copper is oxidized and dissolved. The limpid bead is then re-fused in the R. Fl. with addition of tin. A trace of copper may thus be made to produce distinctly the characteristic reaction, rendering the cold bead distinctly red and wholly or partially opaque.

§ 72. To show the presence of copper in compounds which contain much nickel, cobalt, iron, and arsenic, the assay [No. 82] is first treated with Bx. on Ch. in the R. Fl., when the greater part of iron and cobalt are dissolved. The remaining globule is then mixed with some pure lead, and treated as shown § 71. Arsenic is for the most part driven off, and the rest of the iron and cobalt, with some nickel, absorbed by the boracic acid. The globule is removed from the glass and treated with S. Ph. in the O. Fl.; dark-green while hot, and somewhat lighter green when cold (produced by the mixture of the yellow of nickel and the blue of copper), indicates the presence of copper.

To detect copper when in combination with tin v. § 110.

§ 73. To detect copper in sulphides, the pulverized assay [No. 76] is calcined, and the calcined mass treated as above, or, when the amount of copper is not very small, simply treated with Bx. or S. Ph. on Ch. in the O. Fl., and subsequently with addition of tin in the R. Fl. The presence of copper is then shown by the red color and the opaqueness of the glass on cooling. This reaction is only prevented, or at least made indistinct, by antimony or bismuth, which cause the glass to turn gray or black. In this case the assay is, after calcination, mixed with Sd. Bx., and some pure lead, and the mixture fused on Ch. in the R. Fl. The metallic globule is then heated on Ch. to drive off the antimony, and afterwards treated with boracic acid as above.

§ 74. When a mineral which contains copper is heated in the blue cone, the outer cone of the flame frequently assumes a green, or if the metal is in combination with chlorine, an azure-blue color. This reaction, if not produced by heating the substance alone, may sometimes be elicited by adding a drop of concentrated hydrochloric acid to the pulverized assay [No. 73], evaporating to dryness, mixing the dry powder with a little water to a stiff paste, fastening this into the hook of a platinum wire, and then exposing it to the blue cone of the flame.

§ **75. Fluorine.** To detect fluorine in those minerals where it occurs only as an accessory element in combination with weak bases, and which at the same time contain water, a small piece of the substance [No. 6] is placed into a glass tube sealed at one end, a wet Brazil-wood paper introduced into the open end, and heat applied. Fluoride of silicon and hydrofluoric acid are evolved; the former is decomposed by the watery vapor, and deposits a ring of silica not far distant from the assay, and the latter turns the red color of the test-paper to straw-yellow. Mica, containing not more than $\frac{3}{4}$ per cent. of fluorine, shows the reaction very distinctly.

§ 76. To show the presence of fluorine in minerals where it is united with strong bases, the finely powdered assay [No. 6] is mixed with one to four parts of bisulphate of potassa and introduced into a glass tube, sealed at one end. Heat is applied until sulphuric acid begins to escape. The sides of the tube become covered with silicic acid, resulting from the decomposition of the gaseous fluoride of silicon. The tube is cut off close above the fused mass, cleaned with water, and carefully dried with blotting paper. The dulled appearance of the glass indicates the presence of fluorine.

§ 77. Another process, and by which the presence of fluorine in all kinds of compounds may be shown, is to mix the pulverized assay with some S. Ph. which has previously been fused on Ch. and then reduced to powder; to place the mixture on platinum foil, which is connected with an open glass tube in such a manner as to constitute a kind of tubular continuation to the former, and to heat with the blow-pipe flame until the mass enters into fusion. If the flame is so directed that the products of decomposition are made to pass through the glass tube, and a moistened Brazil-wood paper is introduced into the other end, the presence of hydrofluoric acid is indicated by the change of color which the latter experiences, and often by its pungent odor. In some cases the glass will also be dulled, or a deposit of silicic acid be formed. This test is very delicate.

§ **78. Gold.** When gold is in combination with metals which are volatile at a high temperature, ex. gr. tellurium, mercury, antimony, it is only necessary to heat the alloy on Ch. with the O. Fl., when the gold remains behind in a pure state, and may be recognized by its physical properties. Lead is removed by the process of cupellation, as explained in § 103.

§ 79. When associated with copper, the presence of which is easily detected by S. Ph. on Ch., the alloy, for example gold-coin, is dissolved in pure melted lead, and the new compound subjected to the process of cupellation on bone-ash. Copper is by this means entirely removed. To test the remaining globule for silver, it is treated with S. Ph. on Ch. in the O. Fl.; the silver is gradually oxidized and dissolved by the glass, which when cold assumes an opal-like appearance. To determine approximately the relative proportions of the two metals, the metallic globule is taken from the cupel, placed in a small porcelain dish containing some nitric acid, and heat applied. If the alloy contains twenty-five per cent. of gold or less, it turns black, the silver is gradually dissolved, and the gold remains behind as a brown or black spongy or pulverulent mass. If the alloy contains more than twenty-five per cent. of gold, the globule turns also black, but the silver is not dissolved. If both metals are present in about equal proportions, the globule remains unaltered. If the amount of gold is considerable, it is indicated by the color of the alloy.

In both of the latter cases it must be fused on coal with borax and at least twice its weight of silver, free from gold, and then treated with nitric acid, when the separation will be complete. To form a gold button, it must be well washed with distilled water and fused on coal with borax, and it will then have the pure gold color and bright surface.

§ 80. When associated with metals, which *per se* are infusible before the blow-pipe, as ex. gr. platinum, iridium, palladium, the metallic globule obtained by cupellation shows much less fusibility than pure gold. The exact nature of the foreign metals cannot be ascertained before the Blp.; the humid way must be resorted to.

§ **81. Iodine.** Iodides, tested with a S. Ph. bead which is saturated with oxide of copper, as shown § 65, impart to the outer flame a fine green color [No. 17].

Fused with bisulphate of potassa in a glass tube closed at one end, violet vapors are evolved, iodine sublimes, and sulphurous acid escapes.

To detect a small amount of iodine in mineral water

which has been freed from the greater part of its chloride of sodium by evaporation, a solution of starch in boiling water, with chlorine water, is generally employed, a blue insoluble compound being thus formed if iodine is present. A few drops of nitric acid may be used in place of the chlorine water.

§ 82. Another method, which is said to surpass in delicacy even the reaction with starch, is to mix the substance with a little air-slaked lime, or a mixture of carbonate of lime and quicklime, to dry the mass until no trace of moisture remains, to add some protochloride of mercury (corrosive sublimate), to rub the whole well together, and to place it in a glass tube closed at one end. The tube is then narrowly drawn out a little above the assay, and the mass heated to redness. Protiodide of mercury is formed, which sublimes in yellow or red needles into the narrow tube. This reaction is founded on the property of lime to decompose the protochloride of mercury, but not the protiodide.

Iron. — The reactions of the oxides of iron, see Table II., 13.

§ 83. The colors which iron imparts to the various fluxes are sufficiently characteristic to ascertain its presence in those metallic compounds which contain no easily fusible substances, by simply treating the assay with Bx. on Ch. in the O. Fl. When lead, tin, bismuth, antimony, or zinc are present, the R. Fl. is employed, and directed in such a manner that it principally touches the glass. Thus, the oxidation and consequent saturation of the bead with the oxides of these metals, is to a great extent prevented. In either case the glass, while still soft, is removed from the globule and exposed on another place of the Ch. to the R. Fl. Those metals the oxides of which are easily reduced, are now precipitated, and the characteristic bottle-green color of iron is clearly observable, unless cobalt be present. In this case the glass is again softened with the R. Fl., separated from the precipitated metals, fastened into the hook of a platinum wire, and treated with the O. Fl. until the whole of the iron may be supposed to be converted into sesquioxide. The glass, while hot, will appear green, and blue when cold, if only a trace of iron is present. But when the amount of iron is more considerable, it will be dark-green while hot and bright-green when cold, the latter color resulting from the mixture of the blue of the cobalt and the yellow of iron. The metals remaining behind on Ch. after the treatment with Bx., and which frequently are only copper and nickel (lead, antimony, and bismuth being volatilized), may be treated as shown § 71.

To detect iron in arsenides and sulphides, the assay is well calcined, and the calcined mass treated as above [No. 86 and No. 79].

Or, the substance is pulverized, mixed with test-lead and borax, and fused on coal in the reducing flame until the glass is colored by the easily oxidizable, non-volatile metals present. At first the whole is covered with the R. Fl., but as soon as the borax has united to one globule, the flame is directed upon this alone, allowing the air free access to the fusing metal. When the fusion is complete, the glass is quickly raised with the forceps from the fluid lead, and, after being treated alone on coal in the R. Fl., to reduce any oxide of lead in it, is tested on platinum wire in the O. Fl. If too dark, it is diluted with borax until it is transparent. After separating the bead by means of boracic acid, the other metals combined with it can easily be recognized by the glass fluxes.

Compounds which fuse easily alone on coal can be treated in the R. Fl. with borax, omitting the test-lead. For example, a small trace of iron may be found in galenite, especially if the glass is further treated with tin.

§ 84. The oxides of iron, when associated with a large quantity of manganese [No. 84 and No. 69], color the Bx. bead on platinum wire in the O. Fl. red. To show the presence of iron the bead is removed from the wire, placed on Ch., and treated with tin in the R. Fl. The vitriol-green color of iron will appear in its purity. When associated with the oxides of manganese and cobalt, a minute quantity of iron cannot very well be detected by means of the blow-pipe alone. When accompanied by the oxides of copper and nickel [No. 78 or 85], the assay is dissolved in Bx. on Ch. in the O. Fl., and the glass treated as shown § 83.

§ 85. The presence of chromium prevents any conclusive deduction as to the presence of iron from the color of the beads. In such a case the substance [No. 71] may be mixed with three parts of nitre and one of Sd., and the mixture fused in small portions into the hook of a thick platinum wire. The alkaline chromate is dissolved in water and the residue treated with the fluxes. The presence of the oxides of iron, when associated with the oxides of uranium, cannot be ascertained by means of the blow-pipe alone.

Lead. — The reactions of lead and its compounds, see §§ 15, 23, 36, and Table II., 15.

§ 86. An alloy of lead and zinc [No. 50] deposits a Ct. of oxide of lead mixed with oxide of zinc; the presence of lead is shown by the color of the Ct. and by the azure-blue tinge which it imparts to the R. Fl. (v. § 23).

An alloy of lead and bismuth [No. 49] deposits a Ct.

somewhat darker than that of pure lead, in which the presence of bismuth may be detected as shown § 59, and the presence of lead by the azure-blue color of the R. Fl.

§ 87. To detect lead in sulphides, the substance is placed on Ch. and treated with the R. Fl.; the lead is detected by its Ct. An admixture of antimony cannot by this means be ascertained, since the ring of sulphate of lead surrounding that of the oxide bears a striking resemblance to the Ct. formed by oxide of antimony. In this case the pulverized assay [No. 85] is mixed with a sufficient quantity of Sd., and treated for a short time with the R. Fl. If no antimony is present a pure yellow Ct. with bluish-white edges is formed; but in presence of antimony this Ct. is surrounded by another white one of oxide of antimony. The oxide of lead Ct. appears, moreover, darker than usual, resembling that of bismuth, owing probably to the formation of antimonate of lead. If this Ct. is scraped off from the Ch., and treated with S. Ph. as mentioned § 59, in the case of bismuth, the bead, on cooling, assumes a black color, whereby, in the absence of bismuth, the presence of antimony is proved. A very small quantity of antimony by this method cannot be found out with certainty, since, by keeping up the blast for some time, the sulphide of sodium begins to vaporize and to coat the Ch. with a ring of sulphate of soda (v. § 30).

§ 88. When sulphide of lead is associated with a considerable quantity of sulphide of copper [No. 89], the metallic globule obtained by the process of reduction does not betray, by its physical properties, the presence of lead. But if the alloy is removed from the flux and played upon with a powerful O. Fl., the greater part of the lead will be volatilized and deposit a Ct.

Chloride of lead before the Blp. first fuses and then

gives two coats; one of the chloride, white and volatile, and another of the oxide less volatile. It also imparts a blue color to the R. Fl.

Phosphate of lead alone on coal, fuses to a globule, and affords no coat, or a very slight one.

§ **89. Lithia.**—To detect lithia in silicates which contain only little of it, proceed as follows: The substance [No. 67] is reduced to a fine powder and mixed with about two parts of a mixture of one part of fluor-spar with one and a half parts of bisulphate of potassa; a few drops of water are added and the whole kneaded into a paste. According to Merlet, it is often necessary to use two parts of the mixture with one part of the assay. The mass is fused with the blue cone of the flame into the hook of a platinum wire. If lithia is present the outer flame will appear red. The color is not very intense, and verging into violet. The presence of potassa does not prevent the reaction, but makes the flame appear still more violet; soda makes the reaction uncertain.

If boracic acid be present in the silicate, as in tourmaline, the outer flame at first exhibits a green tinge, but afterwards a wine or less intense red from the lithia.

A mixture of two parts of ignited gypsum, and one part of fluor-spar, is said to be better than the above.

Another method of detecting lithia when mixed with soda is to dip the assay, moistened with hydrochloric acid, into melted wax and then heat it in the blue flame, by which the red color is produced immediately.

Lithia may also be detected by using the indigo prism and Bunsen burner. The assay powder is heated with gypsum in the zone of fusion, and opposite to it some carbonate of potassa, while both flames are observed through the prism which is passed before the eye. If the assay contains lithia, its flame at that point of the prism where the soda flame disappears seems red in comparison with the corn-flower blue potassa flame. As the stratum increases in thickness the red lithia flame gradually loses its intensity, while the blue of the potassa flame passes through violet to red, which at a certain thickness of the stratum is quite similar to the color of the lithia flame.

Manganese.—The reactions of manganese, see Table II., 16.

§ 90. The presence of manganese in any compound substance is readily detected by mixing the pulverized assay [No. 66 or No. 84] with about two or three parts of Sd., and fusing it by means of the O. Fl. on platinum foil. Manganate of soda is formed, which, while hot, is green and transparent, and, on cooling, turns bluish-green and opaque. The reaction is very distinct when as much as one-tenth per cent. of manganese is present. But even the slightest trace may be detected when, instead of Sd., a mixture of one part of nitre with two parts of Sd. is used. Chromium does not prevent the reaction, merely changing the color to yellowish-green. It is only in presence of silica and cobalt that this test is not available, since at a high temperature the silica unites with the soda to silicate of soda, which, in dissolving the oxide of cobalt, produces a blue glass, and thus interferes with the manganese color.

Mercury.—The reactions of mercury and its compounds, see §§ 11, 17, and Table II., 17.

§ 91. Mercury is detected in amalgams [No. 47] by the sublimate of metallic mercury which they yield, when in a glass tube closed at one end.

When in combination with sulphur [No. 81], chlorine [No. 39], iodine, or oxygen-acids, the substance is previously mixed with some anhydrous Sd. or some neutral oxalate of potassa. The acids, etc., are retained by the soda, and mercury sublimes.

If the quantity of mercury is so small, that the nature of the sublimate cannot with certainty be ascertained, the experiment has to be repeated, a piece of iron wire around which a gold-leaf has been wrapped being at the same time introduced into the tube and held close above the assay. The gold-leaf will turn white, even when the amount of mercury present is very small.

Nickel. — The reactions of nickel, see Table II., 19.

§ 92. To detect nickel in metallic compounds which are fusible before the Blp., the assay is treated with Bx. on Ch. in the R. Fl.; iron, cobalt, etc., enter into the flux and may be detected as shown § 69, while the metals the oxides of which are easily reduced remain behind. This operation is repeated until the glass appears no longer colored. The remaining globule is treated with S. Ph. in the O. Fl. We now obtain either the pure color of nickel, or that of nickel mixed with copper, yellowish-green (v. § 72); in this case it is treated on Ch. with tin, whereby the presence of copper may be ascertained, the bead becoming opaque and red. Bismuth or antimony prevents the reaction for copper, the bead turning black instead of red. Such compounds must, previous to their treatment with fluxes, be heated on Ch. in R. Fl. until all volatile substances are driven off [No. 82].

In arsenides and sulphides nickel is detected by the methods given for cobalt under the same circumstances (v. § 70).

§ **93. Nitric Acid.**—The perfectly dry substance [No. 23] is heated in a matrass with some bisulphate of potassa; orange-yellow vapors of nitrous acid are emitted, even if but a small quantity of a nitrate is present. Or the substance may be heated with litharge, free from peroxide of lead, which at first absorbs the nitric acid, but yields it up at a higher temperature. A piece of paper moistened with a solution of protosulphate of iron, free from sesquioxide and acidulated with sulphuric acid, is inserted into the neck of the tube, which should be rather long, and nitrous acid if present will color the paper yellowish to brown.

§ **94. Phosphoric Acid.** — A very minute quantity of phosphoric acid may be detected by pulverizing the substance [No. 14], adding a drop of concentrated sulphuric acid, fastening the paste into the hook of a platinum wire, and playing upon it with the blue cone of the flame; the outer flame will assume a bluish-green color (v. § 35).

Certain nitrogen compounds, as nitric acid, nitrate of ammonia, chloride of ammonium, etc., when fastened into the hook of a platinum wire and touched with the cone of the blue flame, impart to the outer flame a bluish-green color, resembling that caused by phosphoric acid.

§ 95. In a substance, containing not less than about five per cent. of phosphoric acid, the presence of the latter may be shown by dissolving the assay [No. 68] on Ch. in vitrified boracic acid, and forcing into the glass, when a good fusion is effected, a piece of fine steel wire; a good R. Fl. is then given. The iron is oxidized at the expense of the phosphoric acid, causing the formation of a borate of the oxide of iron, which fuses at a sufficiently high temperature. The bead is then taken from the Ch., enveloped in a piece of paper, and struck lightly with a hammer, by which means the phosphide of iron is separated from the surrounding flux. It exists as a metallic-looking button, attractable by the magnet, frangible on the anvil, the fracture having the color of iron. If the substance under assay contained no phosphoric acid, the iron wire will keep its form and metallic lustre, excepting at the ends, where it will be oxidized and burnt. The substance to be assayed ought not to contain sulphuric acid, arsenic acid, or any metallic oxides reducible by iron; but if so, a preliminary test should be made for them.

§ 96. Phosphate of lead, owing to the lead, gives a blue flame, the tip of which has a green color, and also exhibits the peculiarity of crystallizing on cooling after having been fused on Ch; the crystals have frequently large facets of a pearly lustre.

For very small quantities of phosphoric acid, Bunsen has proposed a test which consists in mixing the substance with two or three times as much soda, and placing the completely dried mixture in the drawn-out part of a small tube, similar to those used in testing for arsenic. The mixture is again heated to remove all moisture, a long bit of sodium inserted into it, and fused with the Blp. When cold, the portion of the tube containing the fused mass is broken off, placed in a porcelain dish, and wet with a few drops of water; if phosphoric acid was present, the phosphuretted hydrogen formed may be recognized by its odor.

§ **97. Potassa.** — The violet color of the flame is sufficiently characteristic for potassa (v. § 33). But being altogether prevented, or at least made very indistinct by the addition of a few per cent. of soda or lithia, it can only in a very few cases be made use of. For the detection of potassa in silicates it is almost entirely unavailable, because these compounds almost always contain some soda.

§ 98. If the base of a compound consists essentially of potassa, the following method may be advantageously employed for its detection: Some Bx. to which a little boracic acid has been added, is melted into the hook of a platinum wire, and so much pure protoxide of nickel, free from cobalt, added that the glass on cooling shows a distinct brownish color. A small piece of the substance under examination [No. 15] is made to adhere to the glass and the whole fused together with the O. Fl. If the assay-piece contained no potassa, the color of the glass, after perfect cooling, will have remained unchanged; but if potassa was present in sufficient quantity, the glass will appear bluish.

The simplest means of detecting potassa in a salt, in which, owing to a greater or less amount of soda, the violet coloration of the flame cannot be recognized, consists in viewing the color of the flame through deep-blue cobalt glass, or a stratum of indigo solution, see page 213. The presence of potassa is recognized, according to the thickness of the intervening medium, by the violet or poppy-red color, while a very large amount of soda produces a blue color, and a smaller quantity is not perceptible.

The carbon of organic matter produces the same color as potassa, and if contained in the assay should be removed by ignition.

If lithia is present, a thicker stratum of solution or darker glass must be used.

According to Merz, a green glass may be used in some cases with advantage, the lithia flame being invisible through it, while the potassa and baryta flames appear bluish-green, and that of soda orange-yellow.

In testing silicates with the cobalt glass, they should first be heated with pure gypsum in the flame, thus forming sulphate of the alkali, which is volatile and colors the flame.

99. Selenium.—The reactions of selenium are very characteristic. In non-volatile compounds, which do not give the red sublimate mentioned § 11, the selenium is detected by heating a small piece of the substance [No. 87] on Ch. in O. Fl., when the peculiar odor is evolved; if much selenium is present, a Ct. is deposited, v. § 28. Selenites and selenates are treated on Ch. with Sd. in R. Fl., when a reduction takes place and the selenium vaporizes with the characteristic odor.

§ **100. Silica.**—Pure silica [No. 54], when treated with Bx. on platinum wire, dissolves slowly to a transparent glass which fuses with difficulty. Treated with S. Ph. in the same manner, only a small quantity is dissolved, the rest floating in the liquid bead as a semi-transparent mass. The behavior to Sd., see § 39. With a little So. Co. it assumes a pale-bluish color, which, on addition of a large quantity of the reagent, turns dark-gray or black; very thin splinters may be fused by a great heat to a reddish-blue glass.

§ 101. Silicates [No. 61], when treated with S. Ph. on platinum wire, are decomposed; the bases unite with the free phosphoric acid to a transparent glass, in which the silica may be seen floating as a gelatinous, cloudy mass. The bead ought to be carefully observed while hot, since many silicates form a glass which on cooling opalizes or becomes opaque, when, of course, the phenomenon can no longer be seen. The experiment is best performed with a small splinter of the substance under examination, and only when this does not appear to be affected by the flux, the finely pulverized substance should be used. If but a very small quantity of silica is present, the glass will appear perfectly transparent. Its presence in this case cannot be detected by means of the Blp.

§ 102. Silicates containing at least so much silica that

the quantity of oxygen in the acid is twice that of the oxygen in the base, dissolve, when treated with Sd. on Ch., with effervescence to a transparent glass, which remains so when cold. When less silica is present decomposition also takes place, but the glass turns opaque on cooling, the amount of silicate of soda which is formed not being sufficient to dissolve the eliminated bases.

Silver. — The reactions of silver, see § 27, and Table II., 27.

§ 103. When in combination with metals which are volatile at a high temperature, ex. gr. bismuth, lead, zinc, antimony, the substance is heated alone on Ch., when, after volatilization of these metals by long blowing, a button of pure silver remains behind, and a reddish Ct. is deposited on the Ch. If associated with much lead or bismuth, these metals are best removed by cupellation, a process which is executed in the following manner: Finely pulverized bone-ash is mixed with a minute quantity of soda, and made with a little water into a stiff paste; a hole is now bored into the Ch., filled with the paste, and its surface smoothed and made slightly concave by pressing on it with the pestle of the little agate mortar. The mass is then dried by the flame of a gas or spirit lamp. On this little cupel the assay [No. 51] is placed, and heated with the O. Fl. until the whole of the lead or bismuth is oxidized and absorbed by the cupel. The silver, or if gold is present, the alloy of silver and gold, remains as a bright metallic button on the cupel.

§ 104. When combined with metals which are not volatile, but which are more easily oxidized than silver, the presence of this metal may in some cases be detected by simply treating the alloy with Bx. or S. Ph. on Ch. Copper, nickel, cobalt, etc., are oxidized and their oxides dissolved by the flux, while silver remains behind with a bright metallic surface. But when these metals are present to a considerable extent, another course has to be pursued, a course which may always be taken when a substance is to be assayed for silver, or silver and gold.

§ 105. The assay-piece [No. 86] is reduced to a fine powder, mixed with vitrified Bx. and metallic lead (the quantities of which altogether depend upon the nature of the substance, and for which, therefore, no general rule can be given), and the mass placed in a cylindrical hole of the Ch. A powerful R. Fl. is given until the metals have united to a button, and the slag appears free from metallic globules. The flame is now converted into an O. Fl. and directed principally upon the button. Sulphur, arsenic, antimony, and other very volatile substances, are volatilized; iron, tin, cobalt, and a little copper and nickel become oxidized and are absorbed by the flux; silver and gold and the greater part of the copper and nickel remain with the lead (and bismuth, if present). When all volatile substances are driven off, the lead begins to become oxidized and the button assumes a rotary motion; at this period the blast is discontinued, the assay is allowed to cool, and when perfectly cold the lead button is separated from the glass by some slight strokes with a hammer. It is now placed on a cupel of bone-ash and treated with the O. Fl. until it again assumes a rotatory motion. If much copper or nickel is present, the globule becomes covered with a thick infusible crust, which prevents the aimed-at oxidation; in this case another small piece of pure lead has to be added. The blast is kept up until the whole of the lead and other foreign metals, viz., copper and nickel, are oxidized; this is indicated by the cessation of the rotatory movement, if only little silver is present, or by the appearance of all the tints of the rainbow over the whole surface of the button, if the ore was very rich in silver; after a few moments it takes the look of pure silver. The oxides of lead, copper, etc., are absorbed by the bone-ash, and pure silver, or an alloy of silver with other noble metals, remains behind; the button may be tested for gold, etc., after the method given in § 79.

The chloride of silver can be reduced on coal with soda.

§ **106. Sulphur.**— The presence of sulphur in sulphides may in many cases be detected by heating in a glass tube (v. §§ 11, 14), or on Ch. with the O. Fl.

§ 107. A very delicate test for the presence of sulphur, in whatever combination it may be contained in the substance, and which possesses moreover the advantage over all other methods of being very easily performed, is to mix the pulverized assay [No. 4] with some pure Sd., or, better still, with a mixture of two parts of Sd., perfectly free from sulphates, and one of Bx., and to treat it on Ch. with the R. Fl. The fused mass is removed from the Ch., powdered, the powder placed on a silver foil or a bright silver coin, and a drop of water added. If the substance under examination contained any sulphur, a black spot will be formed on the silver foil, owing to the formation of sulphide of silver from the decomposition of the sulphide of sodium, which, in its turn, resulted from the decomposition of the sulphide or sulphate, or other sulphur-compound of the assay-piece, under the influence of Sd., Ch., and a high temperature. Selenium shows the same reaction, but is readily recognized by the peculiar odor which it emits when heated on Ch. alone.

Dana suggests the following test. Fuse the substance with soda in the R. Fl., moisten with a drop of water in a watch glass, and add a bit of nitro-prusside of sodium. If sulphur in any form is present a purple color will be produced.

A dilute solution of molybdate of ammonia with an excess of hydrochloric acid is colored fine blue by a small quantity of sulphuretted hydrogen or sulphides dissolved in water.

To decide whether the reactions obtained in the experiments above were owing to the presence of a sulphide or to that of a sulphate, the finely-pulverized substance [No. 79] is fused in a small platinum spoon with some hydrate of potassa. The spoon with the contents is then placed in a vessel containing some water, and a piece of silver foil inserted into the liquid. If the silver remains perfectly bright, a sulphate was present; if it turns black, a sulphide. The absence of substances which might exercise a reducing influence is required.

§ **108. Sulphuric and Sulphurous Acids.** — When the bases of the salts do not color the beads, sulphuric acid can be detected by forming with soda and silica on coal a bead in the R. Fl. which is clear and colorless, which is fused with a little of the salt in the R. Fl., and then observing the color. Sulphuric acid gives a yellow to a dark-red color, according to the amount present. If the bases produce color, the salt must first be decomposed by mixing it with twice its amount of soda, igniting it in the R. Fl. on platinum wire or foil, dissolving the sulphate of soda formed in a few drops of water, evaporating the clear solution to dryness on platinum foil or in a small porcelain dish, and testing the salt as above with silicate of soda.

Or mix the substance to be tested with two or three times its weight of soda, or neutral oxalate of potassa (quite free from sulphate), and fuse in the R. Fl. on coal. The mass, with a piece of the coal into which it has sunk, is placed on a bright silver foil and moistened with water, when a brown or black stain will appear. The fused mass may also be covered with hydrochloric acid in a matrass, and the presence of sulphuretted hydrogen tested by inserting a strip of filter paper, moistened with a solution of acetate of lead, into the mouth of the matrass. The paper will be colored brown or black.

§ **109. Tellurium.** — The presence of tellurium in mineral substances is detected by the tests given §§ 11, 18, 29. In presence of lead or bismuth, the reactions in the open tubes and on Ch. are not quite pure. In this case we may subject the assay to the following treatment: The substance is mixed with some Sd. and charcoal-powder, the mixture introduced into a glass tube closed at one end, and heated to fusion; after cooling, a few drops of hot water are poured into the tube; if tellurium was present, telluride of sodium has been formed, which dissolves in hot water with a purplish-red color. This test is applicable to show the presence of tellurium in a great many compounds, even when it occurs in the oxidized state.

Natural tellurium compounds, when gently heated in a matrass with an excess of sulphuric acid, impart to it a purple or hyacinth-red color, which disappears on adding water, while a blackish-gray precipitate is formed. When a mineral containing tellurium is treated on coal it generally yields a white, tellurous acid coat, with a reddish-yellow border, which disappears under the R. Fl., imparting to the flame a green, or, in presence of selenium, a bluish-green tinge. The horse-radish odor would be a certain indication of selenium.

If the mineral contains lead or bismuth, and is treated alone on coal for only a few moments, no pure tellurous acid coat is obtained, but a mixture of this with oxide of lead or bismuth is liable to be deposited. This difficulty can be remedied by mixing the powdered assay with an equal volume of vitrified boracic acid and treating it in the R. Fl. The oxide of lead or bismuth is dissolved in the boracic acid, notwithstanding the reducing flame, and yields no coat, while the tellurium alone volatilizes and coats the coal. When much selenium is also present a portion of it is deposited on the coal, and then the tellurous acid coat is less distinct. In such cases the mineral must also be tested in the open tube.

Tin. — The reactions of tin and its compounds, see §§ 12, 26, 45, and Table II., 30.

§ 110. The presence of tin is indicated by its Ct. when the substance [No. 13], alone or mixed with Sd., is exposed to the R. Fl. on Ch.

When the substance under examination is an alloy, a little Bx. is conveniently added, which absorbs the oxide of tin in a measure as it is formed, and allows the presence of those metals which are more volatile, ex. gr. antimony, lead, bismuth, to be recognized by their yellow coatings. In case of doubt as to the presence of lead or bismuth, the coat is carefully scraped off, dissolved in S. Ph. on platinum wire, and the bead treated on coal with tin. Bismuth is shown by the gray or black appearance of the cold bead. Arsenic is detected by its odor, and iron by the color which the Bx. bead assumes when refused on platinum wire in the O. Fl.

To detect copper in tin or its alloy, the assay [No. 52] is fused with a flux consisting of one hundred parts of Sd., fifty of vitrified Bx., and thirty of silica. The flame is so directed that the metallic globule assumes a rotatory motion. When in this state the glass is kept covered as much as possible with the O. Fl., care being taken that the globule is at one side in contact with the glass, and at the other with the Ch. The tin becomes oxidized, and the oxide, in a measure as it is formed, absorbed by the flux; the remaining button is copper, pure, or with a small quantity of tin, and may be readily tested with the usual fluxes.

To detect a little copper in tin it is treated with successive portions of S. Ph. on coal in the O. Fl. until nearly all the tin is separated and the remaining button imparts a bluish-green color to the glass, when a bit of pure tin is added and the glass treated a short time in the R. Fl.; on cooling, the bead becomes red.

Tin, when present in alloys, is almost always detected on fusing them upon coal; the globule is crusted with oxide, which can be removed with some difficulty after adding borax.

Sulphides containing tin, but forming no coat of oxide of tin near the assay, when treated alone on coal, must be roasted and treated in the R. Fl. with soda and borax, when metallic tin is obtained, which may be tested alone on coal. If other reducible metals are present they form an alloy, in which the other metals can be recognized by means of the fluxes.

In metallic oxides, or substances generally composed of oxides, tin may be detected by a reduction assay on coal with soda or neutral oxalate of potassa. It is sometimes necessary to add borax to absorb the iron.

§ **III. Titanium.** — Titanic acid, when forming the principal constituent of any mineral substance, is easily detected by its behavior with the fluxes, v. Table II., 31; but when in combination with bases these reactions are not always clearly perceptible, being frequently obscured by the predominating reaction of the base. In such cases we may subject the assay to the following treatment, by which even very small quantities of titanic acid will become apparent: the substance [No. 65] is reduced to a very fine powder, mixed with from six to eight parts of bisulphate of potassa, and fused in a platinum spoon at a low red-heat; the fused mass is dissolved in a porcelain vessel in the smallest possible quantity of water, aided by heat. If concentrated, may be heated to boiling. There remains an insoluble residue, which is allowed to settle; the clear liquid is poured off into a larger vessel, mixed with a few drops of nitric acid and at least six volumes of water, and heated to ebullition. If the substance under examination contained any titanium, a white precipitate of titanic acid forms on boiling. If the solution is not acidified with nitric acid before boiling, a yellow, ferruginous titanic acid is obtained when the substance contains iron. The precipitate is collected on a filter, washed with water, acidulated with nitric acid, and tested with S. Ph. either on platinum wire or on coal. If the amount of titanic acid is so small that it does not give in the R. Fl. to the S. Ph. the violet color of sesquioxide of titanium, it is only necessary to add a little sesquioxide of iron when the assay is upon a wire, or a small piece of iron wire when on coal, and to fuse the glass for a short time with the R. Fl.; it appears yellowish while hot and brownish-red when cool.

§ **112. Uranium.** — The presence of this metal is easily recognized, in substances which contain no other coloring constituents, by the reactions given Table II., 33; the most characteristic test is that with S. Ph. In presence of much iron this reaction becomes indistinct; we may then operate in the following manner: the finely-pulverized substance [No. 70] is fused with bisulphate of potassa, the fused mass dissolved in water, mixed with carbonate of ammonia in excess, the liquid separated from the precipitate by filtration, and the filtrate heated to ebullition. If any uranium was present, a yellow precipitate is thrown down, which gives with the fluxes the reactions of pure uranium.

If the substance contains oxide of copper, it is treated with soda borax and a silver button on coal in the R. Fl. until all the copper is reduced into the silver, after which the slag, containing uranium and other non-reducible oxides, like oxide of iron in a low state of oxidation, is dissolved by warming it with a little aqua regia treated with excess of carbonate of ammonia, and the process conducted as above.

Zinc. — The reactions for zinc and its compounds, see §§ 12, 25, 45, and Table II., 35.

§ 113. A small amount of zinc, when associated with considerable quantities of lead, or bismuth, or antimony, or tin, cannot always with certainty be ascertained by means of the Blp.

If the substance under examination contains the zinc as oxide [No. 36], or but a small quantity of sulphide, it is mixed with Sd. and treated on Ch. in R. Fl. Substances consisting essentially of sulphide of zinc may be thus treated without the addition of Sd., and such as contain, besides oxide of zinc, other metallic oxides, are conveniently mixed with some Sd. to which about one-half of its weight of Bx. has been added. A ring of oxide of zinc is deposited on the Ch. When lead is present [No. 50] the Ct. is frequently not pure, being mixed with the Ct. of lead. In this case it is moistened with some So. Co. and heated again with the O. Fl. The oxide of lead is reduced by the red-hot Ch. and volatilized, while the oxide of zinc remains behind with a green color (v. § 45).

In the presence of much antimony a little zinc can only be found with difficulty before the Blp., for the acids of antimony formed will have a green color and cannot be driven off with the O. Fl. In many compounds, however, all of the antimony may be volatilized with the O. Fl. and the zinc then treated with the So. Co.

If tin is present the zinc cannot be recognized by the coating on coal, as its oxide assumes a bluish-green color with So. Co.

CHAPTER IV.

Characteristics of the most important Ores: their behavior before the Blow-Pipe, and to Solvents.

§ 114. Of the physical properties of the minerals which are treated of in this chapter, only those are enumerated which serve best to discriminate the different ores from each other. For a more detailed description, the student must refer to Dana's and other works on mineralogy. Among the distinguishing characters of minerals, their crystalline form, hardness, and specific gravity stand foremost. The latter cannot be ascertained without a good balance, and will, for this reason, be of much less use to the practical man than the determination of hardness, an operation which may be performed in a few moments. A set of minerals, representing the scale of hardness, being not always at hand, it will be useful to give a series of substitutes for them, as arranged by Mr. Chapman:

1. Yields easily to the nail.

2. Yields with difficulty to the nail, or merely receives an impression from it. Does not scratch a copper coin.

3. Scratches a copper coin; but is also scratched by it, being of about the same degree of hardness.

4. Not scratched by a copper coin; does not scratch glass.

5. Scratches glass, though with difficulty, leaving its powder upon it. Yields readily to the knife.

6. Scratches glass easily. Yields with difficulty to the knife.

7. Does not yield to the knife. Yields to the edge of a file, though with difficulty.

8, 9, 10. Harder than flint.

The scale of hardness, as introduced by Mohs, and enlarged by Breithaupt, is as follows:

1. Talc; common laminated light-green variety.
2. Gypsum; crystalline variety.
3. Calcite; transparent variety.
4. Fluorite; crystalline variety.
5. Apatite; transparent variety.
6. Orthoclase; white cleavable variety.
7. Quartz; transparent.
8. Topaz; transparent.
9. Corundum; cleavable varieties.
10. Diamond.

To test the hardness of a mineral we may proceed in two different manners: firstly, by attempting to scratch it with the minerals enumerated in the scale, successively; or, secondly, by abrasion with a file. If the file abrades the mineral under trial with the same ease as No. 4, and produces an equal depth of abrasion with the same force, its hardness is said to be 4. If with more facility than 4, but less than 5, the hardness may be 4.2 or 4.5. Several successive trials should be made to obtain certain results; and, when practicable, both methods should be employed.

For scale of fusibility, see Chap. VI.

ORES OF ANTIMONY.

§ 115. **Stibnite** [Gray Antimony]. — Sb^2S^3. H = 2. G = 4.5. 71.8Sb. Orthorhombic. Of lead-gray color and streak, and metallic lustre. Usually of columnar structure, consisting of a vast number of needle-shaped crystals, sometimes side by side, sometimes divergent. Very brittle.

It fuses readily in the flame of a candle. In a matrass, sometimes yields a slight sublimate of sulphur; on increasing the heat by application of the Blp. flame, a sublimate is produced which after cooling is brownish-red, and which consists of a mixture of tersulphide of antimony with antimonous acid. In an open glass tube, emits sulphurous acid and antimonial fumes. On Ch. it is volatilized, covering the Ch. with oxide of antimony, which, when touched with the R. Fl., disappears with a pale greenish-blue tinge.

When pure, wholly soluble in heated hydrochloric acid with evolution of sulphuretted hydrogen; usually a residue of chloride of lead is left. Partly decomposed by caustic potassa; the solution, when mixed with an acid, affords a yellowish-red precipitate.

§ **116. Berthierite.** — Composition variable, sometimes $FeS + Sb^2S^3$. 57Sb, 13Fe. $H = 2 - 3$. $G = 4 - 4.3$. Orthorhombic. Metallic lustre, less splendent than gray antimony; color, dark steel-gray; surface often covered with iridescent spots.

Heated in a matrass, fuses and yields a slight sublimate of sulphur; on application of a strong heat, a black sublimate of sulphide of antimony is formed, which, on cooling, becomes brownish-red. In an open glass tube it behaves like the preceding ore. In Ch. fuses easily and coats the charcoal with oxide of antimony; there remains, finally, a black slag, which is attracted by the magnet and gives with fluxes the iron reaction.

Soluble in hydrochloric acid.

§ **117. Kermesite** [Red Antimony]. — $2\ SbS^3 + SbO^3$. 75.3Sb. $H = 1 - 1.5$. $G = 4.5 - 4.6$. Monoclinic.

Usually in tufts of capillary crystals of cherry-red color, with adamantine lustre. Streak, brownish-red.

In a matrass, fuses readily and yields a slight yellowish-red sublimate; with strong heat, boils and gives a black sublimate, which when cold is brownish-red. In an open tube and on Ch. behaves like gray antimony.

It dissolves in hydrochloric acid with evolution of sulphuretted hydrogen. The powdered mineral when treated with caustic potassa assumes an ochre-yellow color and dissolves completely.

ORES OF ARSENIC.

§ **118. Native Arsenic.**—As, with traces of Sb., Ag., Fe., Co., and Ni. H = 3.5. G = 5.9. Rhombohedral. Of dull metallic lustre; color and streak, tin-white, tarnishing on exposure to air to dark-gray.

Heated in a matrass, sublimes; on Ch. behaves like pure arsenic. In both cases, sometimes, a residue is left which, when treated with fluxes, exhibits the reactions of iron, cobalt, and nickel. (See § 83.)

§ **119. Realgar.**—AsS. 70.1As. H = 1.5—2. G = 3.4—3.6. Monoclinic. Usually of bright-red, sometimes of orange-yellow color, and resinous lustre. Sectile. Streak, orange.

In a matrass, fuses, boils, and finally sublimes; the sublimate after cooling is red and transparent. In an open glass tube, when carefully heated, yields a sublimate of arsenous oxide, sulphurous acid escaping. On Ch. fuses readily and burns with a yellowish-white flame, emitting grayish-white fumes which possess the peculiar alliaceous odor. Subjected to the treatment described § 55, a sublimate of metallic arsenic is obtained.

Not easily affected by acids; but aqua regia dissolves it

with continued digestion, part of the sulphur being precipitated. A heated solution of caustic potassa decomposes it, leaving a brownish-black powder (As^6S) undissolved.

§ 120. **Orpiment.**—As^2S^3. As61. H=1.5—2. G=3.4. Orthorhombic. A foliaceous mineral of lemon-yellow color and streak, and resinous or pearly lustre. Sectile.

Before the Blp. behaves like the preceding, with this difference, that the sublimate, after cooling, is dark-yellow and transparent.

Soluble in aqua regia, caustic potassa, and ammonia.

§ 121. **Arsenolite** [White Arsenic]. — AsO^3. H=1.5. G = 3.1. Isometric. Occurs usually in minute capillary crystals of a white color and streak, and vitreous or silky lustre.

Before the Blp. it behaves like pure arsenous oxide (v. §§ 11, 15, Table II., 2).

Slightly soluble in hot water; more so in water acidulated with hydrochloric acid.

ORES OF BISMUTH.

§ 122. **Native Bismuth.** — Bi: H = 2 — 2.5. G = 9.7. Hexagonal. Color and streak silver white, tinged with red. Lustre metallic. Brittle when cold; but when hot may be laminated. Occurs foliated, granular, and arborescent; occasionally crystallized.

Before the Blp. it behaves like pure bismuth (v. §§ 17, 22).

Readily dissolved by nitric acid; the solution is precipitated by water.

§ 123. **Tetradymite** [Telluric Bismuth].—Bi. and Te. in variable proportions. 53—61Bi, 34—48Te. H=1.5—2. G = 7.2 — 8.4. Hexagonal. Of pale steel-gray color, and high metallic lustre. Occurs usually in tabular crystals or foliated masses; the laminæ are elastic. It soils paper.

In an open glass tube it fuses readily, emitting a white smoke which partly condenses, coating the tube near the assay-piece with a white powder, intermixed with red spots; on directing the flame on this Ct. it fuses to colorless drops (TeO^2), while the red sublimate (Se) disappears. On Ch. fuses instantly to a metallic globule which, when touched with the inner flame, imparts a bluish-green color to the outer one, sometimes gives out selenium vapors, and deposits, close to the assay-piece, a dark orange Ct. surrounded at a greater distance by a white Ct.

Soluble in nitric acid.

§ **124. Bismutite.** — $3 (BiO^3.CO^2 + HO) + BiO.HO$. $90.1 BiO^3$. $H = 4 - 4.5$. $G = 6.9$. Streak usually of a white or light-greenish color; lustre vitreous; in acicular crystallizations.

In a matrass, decrepitates, yields a little water, and turns gray. On Ch. fuses very readily and is reduced, with effervescence, to a metallic globule, covering the Ch. with a Ct. of oxide of bismuth. If the blast is kept up for some time the whole of the bismuth is volatilized, and there remains a scoriaceous mass which in the R. Fl. may be fused to a globule, and which with fluxes gives the indications of copper and iron. With Sd. it usually gives the sulphur reaction (§ 107).

Dissolves in hydrochloric acid with effervescence; the solution has a yellow color.

§ **125. Bismuthinite.** — BiS^3. $81.25 Bi$. $H = 2 - 2.5$. $G = 6.4 - 6.55$. Orthorhombic. In acicular crystals or massive; of metallic lustre, and lead-gray color, with a yellowish or iridescent tarnish. Streak, shiny.

In a matrass, fuses and yields a slight sublimate of sulphur. Carefully heated in an open tube, it fuses and yields sulphurous acid and a coat of sulphate of bismuth; the latter may be fused, by application of the Blp. flame, to brown drops which, when cold, appear yellow and opaque. On Ch. fuses and boils, throwing out small drops in a state of incandescence, and deposits a Ct. of oxide of bismuth.

Soluble in nitric acid with deposition of sulphur. The solution gives a white precipitate with water.

Chiviatite.— $2CuS, 3BiS^3, + 4 (Pb^2S, Bi^3S^3)$. 62.9Bi. Decomposed with nitric acid, with separation of sulphate of lead.

§ **126. Bismite** [Bismuth Ochre]. — BiO^3, containing minute quantities of Fe^2O^3, CuO, and AsO^5. 89.65Bi. G = 4.36. Occurs usually pulverulent or earthy. Streak, straw-colored.

Before the Blp. it behaves like pure oxide of bismuth. Soluble in nitric acid.

ORES OF CHROMIUM.

§ **127. Chromite** [Chromic Iron].—(FeO, CrO, MgO) + (Cr^2O^3, Al^2O^3). $68Cr^2O^3$. H = 5.5. G = 4.3 — 46. Isometric. Occurs usually massive; of iron-black or brownish-black color, with a shining and somewhat metallic lustre. Some varieties are magnetic. Streak, brown.

Heated in a matrass, remains unchanged. Infusible in the forceps. After having been exposed to the R. Fl. it follows the magnet. In Bx. and S. Ph. slowly, but completely, soluble to a transparent glass, which is beautiful green after cooling. Mixed with Sd. and nitre, and heated on platinum-foil, the mass fuses and becomes yellow. With Sd. on Ch. in R. Fl. it affords metallic iron.

Concentrated acids affect it but little, even when finely

pulverized; they dissolve only a little iron. Fused with caustic potassa, chromate of potassa is formed.

ORES OF COBALT.

§ **128. Smaltite** [Smaltine]. — (Co, Fe, Ni) As^2. H = 3.5 — 6. G = 6.4 — 7.2. Isometric. Of tin-white or steel-gray color, and metallic lustre. Streak, dark brown.

In a matrass, usually yields, when heated to redness, a sublimate of metallic arsenic. In an open glass tube, affords a copious sublimate of crystallized arsenous oxide, and sometimes emits sulphurous acid. On Ch. it fuses readily, with emission of copious arsenical fumes, to a grayish-black magnetic globule which, with the fluxes, gives the indications of iron, cobalt, and nickel.

With nitric acid it gives a pink solution, arsenous oxide being deposited.

§ **129. Cobaltite** [Cobaltine]. — $CoS^2 + CoAs^2$. 35.5Co, 45.2As. H = 5.5. G = 6 — 6.3. Isometric. Of silver-white and sometimes reddish color, and metallic lustre. Streak, gray.

Unchanged in the matrass. In an open glass tube, yields a sublimate of arsenous oxide and vapors of sulphurous acid. On Ch. emits copious arsenical and sulphur fumes and fuses to a dull black metallic globule, which is attracted by the magnet, and which when treated with fluxes gives the indications of cobalt and iron, and sometimes also those of nickel.

Dissolves in heated nitric acid, arsenous oxide being deposited.

§ **130. Linnæite** [Cobalt Pyrites].—$2CoS + CoS^2$. 58Co. H = 5.5. G = 4.8 — 5. Isometric. Of a more or less bright steel-gray color, and metallic lustre. Crystallizes in regular octahedrons. Streak, dark gray.

In an open glass tube, sulphurous acid is abundantly evolved, and sometimes a light sublimate of arsenous oxide formed. On Ch. small pieces of the mineral readily fuse to a globule which when cold is covered with a black rough crust, and which is attracted by the magnet. The pulverized mineral, after having been well calcined, dissolves in Bx. in O. Fl. to a blue transparent bead. In a highly saturated bead of this kind, when treated on Ch. with R. Fl., particles of metallic nickel may be seen floating about.

Soluble in nitric acid, excepting the sulphur.

§ 131. **Erythrite** [Cobalt Bloom].—$3CoO.\ AsO^5+8HO$. $37.55CoO$, $38.43AsO^5$. $H=1.5-2.5$. $G-2.9$. Monoclinic. Usually of crimson or peach-red color; when crystallized, of pearly lustre; frequently dull and earthy, forming incrustations. Streak, pale red.

Heated in a matrass, loses water, and the color changes to blue or green. A small crystal exposed to the inner flame fuses and colors the outer flame pale-blue. On Ch. in R. Fl. emits arsenical fumes, and melts to a dark-gray globule of arsenide of cobalt, which with fluxes gives the pure cobalt-reactions.

Acids dissolve it readily to a rose-colored liquid; the solution in concentrated hydrochloric acid appears blue while hot. The pulverized mineral is partly decomposed by caustic potassa; the powder assumes a bluish-gray color and the solution is sapphire-blue.

Var. A. **Roselite.** — $H=3$. Cleavage distinct and brilliant, lustre vitreous; color, deep rose-red.

Var. B. **Lavendulan.** — $H=2.5-3$. $G=3$. Amorphous, with a greasy lustre; color lavender-blue. Streak, pale blue.

Heated in a matrass, gives out water. In the forceps

fuses easily and colors the outer flame deep-blue; the fused mass becomes crystalline on cooling. On Ch. in R. Fl. it fuses with emission of arsenical fumes. With fluxes, gives the reactions of Co., Ni., and Cu. (see § 92.)

§ 132. **Asbolite** [Earthy Cobalt]. — It is a variety of Wad (see § 184), containing sometimes a considerable quantity of oxide of cobalt (32 per cent.), in combination with silicic or arsenic acid.

With Bx. in O. Fl. gives a dark-violet glass, which in the R. Fl. becomes blue. The S. Ph. bead when treated on Ch. with metallic tin frequently exhibits the copper-reaction. With Sd. on platinum-foil it shows the presence of manganese.

Soluble in hydrochloric acid with evolution of chlorine; the solution is usually blue, and on addition of water becomes red.

§ 133. **Remingtonite.** — It occurs as rose-colored incrustation, soft and earthy; opaque. Streak, pale rose-colored.

It is a hydrous carbonate of cobalt. $CoO.CO^2,HO$. Soluble in hydrochloric acid with effervescence, giving a green solution, the color due to iron. Gives the cobalt-reaction with borax.

It sometimes resembles kermesite, also some varieties of cuprite, but can be easily distinguished from these by the Blp.

ORES OF COPPER.

§ 134. **Native Copper.** — Pure Copper. $H = 2.5 — 3$. $G = 8.9$. Isometric. Of metallic lustre and copper-red color. Occurs usually massive or arborescent.

It fuses on Ch. to a globule which, if the heat is sufficiently high, assumes a bright bluish-green surface; on cooling it becomes covered with a crust of black oxide. With the fluxes it gives the usual indications of copper.

It dissolves readily in nitric acid.

§ 135. **Chalcopyrite** [Copper Pyrites]. — $Cu^2S + Fe^2S^3$. 34.6Cu, 30.5Fe. H = 3.5 — 4. G = 4.1 — 4.3. Tetragonal. Of a brass-yellow color and metallic lustre; on exposure to moist air it becomes iridescent on its surface. It occurs crystallized, but usually massive. It is easily scratched with a knife, giving a greenish-black powder.

Heated in a matrass, decrepitates and yields sometimes a faint sublimate of sulphur, assuming at the same time a darker color or becoming iridescent. Heated in an open glass tube, sulphurous acid is given out abundantly. On Ch. when heated it blackens, but becomes red on cooling; with continued heat it fuses to a black globule, which is attracted by the magnet; this globule is brittle, and reddish-gray in the fracture. The pulverized mineral, after roasting, gives with fluxes the indications of iron and copper. With Sd. on Ch. it is reduced; the metals are obtained in separate masses. Moistened with hydrochloric acid it colors the flame green even previous to fusion.

It dissolves in nitric acid, but more readily in aqua regia, leaving a residue of sulphur.

§ 136. **Bornite** [Purple Copper]. — $(FeCu^2)S$. 50-67Cu, 7-15Fe. H = 3. G = 4.4 — 5. Isometric. When crystalline it usually takes the cubical form, and is of a pale-yellowish color; when massive, its color is copper-red to reddish-brown; it speedily tarnishes, assuming various hues, mostly purple, blue, and reddish. When scratched with a knife it gives a grayish powder. Streak, black.

Before the Blp. it shows pretty much the same behavior as copper pyrites.

Concentrated nitric acid dissolves it, leaving the greater part of the sulphur behind.

§ 137. **Chalcocite** [Copper Glance]. — Cu^2S. 79.8Cu. H = 2.5 — 3. G = 5.5 — 5.8. Orthorhombic. Of a blackish lead-gray color, often with a bluish or greenish tint on its surface. Occurs usually in compact masses, very often shining. Streak, black.

Heated in a matrass, nothing volatile is given out. In an open tube, sulphurous acid is evolved. On Ch. readily fuses to a globule, which boils and emits glowing drops, sulphurous acid escaping abundantly; the outer flame is at the same time colored blue. With Sd. on Ch. it yields a globule of metallic copper.

In heated nitric acid it dissolves, leaving a residue of sulphur.

§ 138. **Tetrahedrite** [Gray Copper]. — $4CuS + Sb^2S^3$, or general formula $4\ (Cu^2, Fe, Zn, Ag, Hg) + (Sb, As, Bi)^2S^3$. H = 3 — 4.5. G = 4.5 — 5. Isometric; tetrahedral. Color and streak between steel-gray and iron-black.

Heated in a matrass, fuses and finally yields a dark-red sublimate of tersulphide of antimony with antimonous oxide. In an open glass tube, fuses and gives thick fumes of antimony (and arsenous oxide) and sulphurous acid; mercury when present condenses in the upper part of the tube, forming a metallic mirror. On Ch. it fuses readily to a globule, emitting thick white fumes and sulphur vapor; coatings of antimonous acid and of oxide of zinc are deposited; the latter is nearer to the assay-piece, and may be tested with So.Co. [v. § 45]. To detect arsenic, v. § 56. To detect mercury, add to the finely-pulverized assay three times its weight of dry Sd. and treat the mixture as directed § 91. The pulverized mineral, after having been well roasted, gives with the fluxes the indications of iron and copper; with Sd. affords metallic copper and a little iron. To detect silver, treat the mineral with pure lead and Bx. as directed § 105.

When pulverized, it is decomposed by nitric acid; the solution has a brownish-green color; antimonous oxide (and arsenous oxide) and sulphur remain undissolved. Caustic potassa effects partial decomposition; the sulphide of antimony (and arsenic) enters into solution, and is, on addition of an acid, re-precipitated.

§ **139. Tennantite.** — 4 $(Cu^2, Fe)S, AsS^3$. 41-51.6Co. H = 3.5 — 4. G = 4.37 — 4.5. Isometric. Always crystallized; metallic lustre; color, blackish lead-gray to iron-black. Streak, reddish-gray.

In a matrass, gives a sublimate of tersulphide of arsenic. In an open tube, sulphurous acid and a sublimate of arsenous acid. On Ch., fuses easily with emission of sulphur and arsenic vapors to a dark-gray globule, which is attracted by the magnet. The pulverized mineral gives, after calcination with fluxes, the reaction of iron and copper.

§ **140. Domeykite** [Arsenical Copper].—Cu^6As. 71.7Cu, 28.3As. H = 3 — 3.5. Reniform, massive, or disseminated; lustre, metallic; color, tin-white to steel-gray; black and soft when impure.

Heated in a matrass, yields a little water and a sublimate of arsenous acid; the assay-piece assumes a silver-white color. In an open tube, affords a crystalline sublimate of arsenous acid. On Ch., fuses easily with emission of a strong alliaceous odor, to a reddish metallic mass, which gives the copper reactions.

Readily soluble in nitric acid; decomposed by hydrochloric acid, metallic arsenic remaining undissolved.

Binnite, $\frac{3}{3}Cu^2S + AsS^2$; **Enargite**, $3Cu^2S + As^2S^3$; **Algodonite**, Cu^2As^2; **Whitneyite**, $Cu^{18}As^2$.

§ **141. Atacamite.** — $3CuO.HO + CuCl.HO$. 14.3Cu. H = 3 — 3.5. G = 4 — 4.3. Orthorhombic. Occurs crystalline, or massive lamellar; color and streak, various shades of bright-green, sometimes blackish-green. Lustre vitreous.

Heated in a matrass, gives out water and a gray sublimate, which, on cooling, becomes grayish-white; the water shows acid reaction. On Ch. fuses readily, colors the outer flame azure-blue, and is finally reduced to a globule of metallic copper; two coatings are deposited on the Ch., the one grayish-white, and the other brownish, which, on being played upon with the R. Fl., change their place with an azure-blue tinge.

Easily soluble in acids.

§ 142. **Cuprite** [Red Copper].—Cu^2O. 88.8Cu. H = 3.5—4. G = 5.8—6. Isometric. Usually of a very intense, deep-red color, occasionally crimson-red; exceedingly friable. Streak, red to brownish.

Heated in the pincers, fuses and colors the outer flame emerald-green; moistened with hydrochloric acid and treated in the same manner, the color is azure-blue. On Ch. it blackens, then fuses quietly, and finally yields a globule of metallic copper which, on cooling, becomes covered with a coating of black oxide.

Dissolves readily in nitric acid. It is also soluble in ammonia: the solution is colorless when the access of air is prevented; on exposure to air it turns blue.

§ 143. **Malachite.**—$2CuO.CO^2 + HO$. 57.4Cu. H = 3.5—5. G = 3.7—4. Monoclinic. Occurs usually in the shape of mammillated concretions; the interior is very compact, and lustre shining, in the fracture sometimes earthy, sometimes silky; of a bright-green color and streak.

Heated in a matrass, gives out water and turns black. On Ch. fuses to a globule, and affords metallic copper when the heat is sufficiently high; heated in the forceps, the outer flame is colored green. With fluxes and Sd. it behaves like oxide of copper (v. Table II., 8).

It dissolves in acids with effervescence; also soluble in ammonia.

§ 144. **Azurite** [Blue Malachite]. — $2(CuO.CO^2) + CuO.HO$. 55.2Cu. H = 3.5 — 4. G = 3.5 — 3.8. Monoclinic. Occurs usually crystallized, or in globular masses of columnar structure. It is easily distinguished by its fine blue color and streak; either earthy or vitreous in lustre.

Before the Blp. and with solvents it behaves like malachite.

§ 145. **Chalcanthite** [Copper Vitriol]. — $CuO.SO^3 + 5HO$. H = 2.5. G = 2.21. Triclinic. Lustre, vitreous; color, various shades of blue; taste, metallic and nauseous.

Heated in a matrass, swells up, gives out water, and becomes white. On Ch., colors the outer flame green, fuses, and affords a button of metallic copper, crusted with a coat of sulphide. After calcination, gives with fluxes the reactions of copper, sometimes also those of iron.

Soluble in water; a polished plate of iron introduced into the solution becomes coated with copper.

§ 146. **Pseudomalachite** [Phosphate of Copper]. — $3CuO.PO^5 + 3(CuO.HO)$, sometimes $2(3CuO.PO^5) + HO + 4(CuO.HO)$. 56.6Cu. H = 4.5 — 5. G = 4 — 4.4. Orthorhombic. Occurs both crystallized and massive. Of adamantine lustre, and dark emerald-green or blackish-green color. Streak, bright green.

In a matrass, gives out water and blackens. A piece previously heated in a matrass fuses in the forceps to a black globule, which becomes crystalline on cooling. With Bx. and S. Ph., behaves like oxide of copper. Strongly heated on Ch. with a sufficient quantity of Sd., nearly all the copper is obtained as a metallic globule. Mixed with an equal volume of metallic lead and fused on Ch., a globule of metallic copper is obtained, surrounded by a fused mass of phosphate of lead, which on cooling crystallizes.

Soluble in nitric acid and in caustic ammonia.

§ **147. Olivenite.** — $3CuO.(AsO^5.PO^5) + CuO.HO$. 45.2Cu. H = 3. G = 4.1 — 4.4. Orthorhombic. Crystallized, or in globular and reniform masses of indistinctly fibrous structure. Color and streak usually olive-green to brown. Lustre, vitreous.

In a matrass, yields a little water. In the forceps, fuses to a globule and colors the outer flame bluish-green; the fused mass crystallizes on cooling. On Ch., fuses with detonation and emission of arsenical vapors to a metallic globule; the globule is white and somewhat brittle, and covered with a brown scoria. Fused with metallic lead, it is decomposed in the same manner as the preceding ore.

Dissolves in nitric acid, also in ammonia.

§ **148. Tyrolite.** — $(5CuO.AsO^5 + 9HO) + CaO.CO^2$. 35.Cu. H = 1 — 2. G = 3. Orthorhombic. Usually reniform, massive; structure, radiate foliaceous. Color, pale-green. Very sectile. Lustre, vitreous.

Heated in a matrass, decrepitates, yields much water, and blackens. On Ch., fuses with emission of arsenical vapors to a gray scoriaceous mass, in which minute globules of metallic copper occasionally appear. When the mineral is fused on Ch., with addition of Sd. and Bx., until the oxide of copper is completely reduced and the slag dissolved in hydrochloric acid, a solution is obtained in which the presence of lime may be shown by the proper reagents. Dissolves in nitric acid with effervescence, also in ammonia.

§ **149. Chrysocolla.** — $CuO.SiO^2 + 2HO$. 45.8CuO. H = 2 — 3. G = 2. Occurs usually as an incrustation. It very much resembles malachite; its color is bluish-green, and it is remarkable for its great compactness; its surface is very smooth, giving it the appearance of an enamel or a well-fused slag. Lustre, vitreous. Streak, when pure, white.

In a matrass, yields water and blackens. In the forceps infusible, coloring the outer flame intensely green. On Ch. in O. Fl. blackens, in R. Fl. turns red. S. Ph. and Bx. dissolve it with the usual indications of copper; the S. Ph. bead shows a cloud of undissolved silica. With Sd. on Ch., affords globules of metallic copper.

It is decomposed by acids, silica remaining undissolved.

ORES OF GOLD, PLATINUM, AND IRIDIUM.

§ **150. Native Gold.** — Combination of Au. and Ag. in variable proportions, sometimes with traces of Fe. and Cu. H = 2.5—3. G = 15.6—19.5. Isometric. Easily distinguished by its malleability, its cutting like lead, its high specific gravity, and its resistance to acids. Color and streak, various shades of gold-yellow. It usually occurs in variously contorted and branched filaments, in scales, in plates, or in small, irregular masses.

On Ch., fuses to a globule which after cooling has a bright metallic surface. With S. Ph. in O. Fl. a bead is formed which opalizes on cooling, or becomes opaque and yellow, according to the amount of silver which it contains.

Resists the action of heated concentrated nitric acid; soluble only in aqua regia.

§ **151. Sylvanite** [Graphic Tellurium]. — $AgAuTe^3$. H = 1.5—2. G = 5.7. Monoclinic. Of metallic lustre and steel-gray color. Very sectile. Streak and color, pure steel-gray to silver-white; sometimes yellow.

In an open glass tube, yields a white sublimate which, when played upon with the flame, fuses to transparent drops. On Ch. fuses to a dark-gray globule, depositing at the same time a white Ct. which, when touched with the R. Fl., disappears, tinging the flame bluish-green (see §§ 29, 35). It finally affords a light-yellow malleable globule of metallic lustre.

Soluble in aqua regia, leaving a residue of chloride of silver. The solution gives a white precipitate with water.

§ 152. **Native Platinum.** — Pt., usually combined with a little Fe., Ir., Os., Pd., Rh., and sometimes Cu. and Pb. H = 4—4.5. G = 16—19. Isometric. Usually occurs in grains of silver-whitish or gray color; malleable and ductile.

Infusible before the Blp. and not acted upon by fluxes. Soluble only in heated aqua regia. The solution gives a yellow granular precipitate with chloride of potassium.

§ 153. **Iridosmine** [Osmium Iridium].—The light variety $IrOs^3$ and $IrOs^4$. H = 6—7. G = 19.3—21.1. Hexagonal. Occurs usually in irregular flattened grains, of metallic lustre and tin-white color; but little malleable.

Infusible before the Blp.; when fused with nitre in a matrass the characteristic osmium odor is produced. The fused mass is soluble in water; the solution gives, on addition of nitric acid, a green precipitate. The dark varieties lose before the Blp. the metallic lustre, and when held in the alcohol flame, impart to it a yellowish-red color and great illuminating power. Not visibly affected by any acid.

ORES OF IRON.

§ 154. **Meteoric Iron.** — Fe., with variable quantities of Ni. (from 1 to 20 per cent.); and traces of Co., Mg., Mn.,

Sn., Cu., Cr., Si., C., Cl., S., and P. H = 4.5. G = 7.3—7.8, rarely as low as 6. Isometric. Lustre, metallic; color, iron-gray; ductile; strongly attracted by the magnet.

Infusible. On Ch. with Bx. or S. Ph. gives only the reactions of iron. To detect the presence of the other heavy metals, the assay-piece must be dissolved in aqua regia, the liquid mixed with ammonia in excess, filtered, and the ammoniacal filtrates precipitated with sulphydrate of ammonia. The precipitate consists of the sulphides of nickel, cobalt, manganese, and copper, which may be collected on a filter and treated with Bx. on Ch. as described § 70.

§ 155. **Limonite** [Brown Hematite]. — $2Fe^2O^3.3HO$. 59.9Fe. H = 5—5.5. G = 3.6—4. Of a dull brownish-yellow color, earthy or semi-metallic in appearance, and often in mammillary or stalactitic forms. Streak, yellowish-brown.

In a matrass, yields water, and red sesquioxide remains; in platinum forceps, fusible on the edges; gives with Bx. and S. Ph. an iron reaction; the clayey varieties treated with S. Ph. give a cloud of undissolved silica; treated with Sd. and nitre on platinum-foil the manganese reaction is almost always obtained.

§ 156. **Hematite** [Specular Iron]. Fe^2O^3. 70Fe. H = 5.5—6.5. G = 4.5—5.3. Rhombohedral. Of a dark steel-gray or iron-black color, and usually of metallic lustre; its powder and streak are red.

Infusible alone; becomes magnetic after roasting, and gives the usual indications of iron with the fluxes; its powder dissolves readily in heated hydrochloric acid. Contains sometimes chromium and titanium, which may be detected by the processes given §§ 68 and 111.

§ **157. Magnetite** [Magnetic Iron Ore]. — FeO. Fe^2O^3. 72.4Fe. H = 5.5—6.5. G = 4.9—5.2. Isometric. Its color is iron-black, with a shining metallic or glimmering lustre; its powder and streak are black; it is strongly attracted by the magnet.

It fuses with difficulty, and gives the usual reactions of iron with the fluxes; the pulverized mineral dissolves completely in hydrochloric acid.

§ **158. Pyrite** [Iron Pyrites]. — FeS^2. 46.7Fe. H = 6—6.5. G = 4.8—5. Isometric. Occurs commonly in cubes. Usually of a brass-yellow color and metallic lustre; sometimes colored or brown by metamorphosis. By its superior hardness, not yielding to the knife, and emitting sparks when struck with steel, it may be distinguished from copper pyrites. Streak, gray.

Heated in a glass tube closed at one end, usually emits some sulphuretted hydrogen, and yields a sublimate of sulphur; the residue is attracted by the magnet. Heated on Ch. with the O. Fl. the sulphur burns off with a blue flame and leaves red oxide behind, which, when treated with the fluxes gives pure iron reactions. But slightly affected by hydrochloric acid; nitric acid dissolves it, leaving a residue of sulphur.

§ **159. Marcasite** [White Iron Pyrites]. — FeS^2. H = 6—6.5. G = 4.6—4.8. Orthorhombic. Crystals are prismatic, often twins. Color usually light bronze-yellow, sometimes inclined to green or gray; occurs frequently in radiated masses or crest-like aggregations. Streak, greenish-gray. Very liable to decomposition.

Before the Blp. it behaves like the preceding.

§ **160. Pyrrhotite** [Magnetic Pyrites]. — 6FeS + FeS^2. 69.5Fe. H = 3.5—4.5. G = 4.4—4.7. Hexagonal. Very much resembles common iron pyrites, from which it is distinguished by its inferior hardness, more reddish color, and by being slightly attracted by the magnet. Streak, dark gray.

Heated in a matrass, remains unchanged; in the open glass tube, emits sulphurous acid but yields no sublimate. On Ch. in R. Fl., fuses to a globule, which is covered with an uneven black coating, which follows the magnet, and which, on a surface of fracture, exhibits a yellowish crystalline structure and metallic lustre. In O. Fl. it is converted into red oxide.

Soluble in hydrochloric acid, excepting the sulphur, with evolution of sulphuretted hydrogen.

§ **161. Arsenopyrite** [Mispickel]. — $FeS^2 + FeAs^2$. 34.4Fe. $H = 5.5—6$. $G = 5—6.4$. Orthorhombic. Of metallic lustre and a silver-white to steel-gray color. Streak, dark grayish-black. Brittle.

Heated in a matrass, yields first a red sublimate of sulphide of arsenic, and afterwards a black crystalline one of metallic arsenic; in an open glass tube, yields arsenous oxide and sulphurous acid. On Ch., emits copious arsenical fumes, and a Ct. of arsenous oxide is deposited; then fuses to a globule which shows the properties of fused magnetic pyrites. Frequently contains cobalt, the presence of which may be detected by the method described in § 69.

Soluble in nitric acid and aqua regia, leaving a residue of sulphur and arsenous oxide; the latter dissolves with continued digestion.

§ **162. Menaccanite** [Titaniferous Iron]. — Ti^2O^3 and Fe^2O^3 in various proportions. $H = 5—6$. $G = 5.5—5$. Rhombohedral. Of iron-black color, usually in tabular crystals; bears a great resemblance to specular iron, but gives no red powder.

Alone in the O. Fl. infusible; in R. Fl. it may be rounded at the edges. With Bx. and S. Ph. in O. Fl., gives the reactions of pure oxide of iron; but the S. Ph. bead when treated with the R. Fl. assumes a brownish-red color, the intensity of which depends upon the amount of titanic acid present; this glass, when treated with tin on Ch., turns violet (v. Table II., § 31). To show conclusively the presence of Ti., follow the method given in § 111.

Dissolved by hydrochloric acid and aqua regia, with separation of titanic acid; some varieties dissolve with great difficulty, even when reduced to a very fine powder.

§ **163. Siderite** [Spathic Iron]. — $FeO.CO^2$. 48.22Fe. H = 3.5—4.5. G = 3.7—3.9. Rhombohedral. Color from grayish-yellow to reddish-brown; crystallizes in rhombohedrons, which are often curved, and are very distinctly cleavable; often massive. Lustre, vitreous; and streak, light brown.

Heated in a matrass, frequently decrepitates, carbonic acid and carbonic oxide are given out, and a black oxide of iron remains, which is attracted by the magnet. Alone, infusible. With Bx. and S. Ph. it gives the pure iron reactions, and with Sd. sometimes those of manganese. It dissolves in strong acids with effervescence, but with difficulty, and only when pulverized.

§ **164. Melanterite** [Copperas]. — $FeO.\ SO^3 + 7HO$. 25.9FeO. H = 2. G = 1.83. Monoclinic. Occurs usually massive and pulverulent, of various shades of green, becoming yellowish on exposure to air; taste, astringent and metallic.

In a matrass, gives out sulphurous acid and water, which shows acid reaction. Strongly heated, only sesquioxide of iron remains. Soluble in water.

§ **165. Vivianite** [Blue Iron Earth]. — $6(3FeO,PO^5 + 8HO) + (3Fe^2O^3,2PO^5 + 8HO)$. H = 1.5—2. G = 2.66. Monoclinic. Occurs crystallized, or in reniform and globular masses, fibrous and radiated, sometimes as incrustation. Color, blue to green, usually dirty-blue. Vitreous lustre and bluish streak.

In a matrass, swells and gives pure water. In the forceps, fuses to a steel-gray metallic globule, coloring the outer flame bluish-green. With fluxes gives the reactions of iron.

Easily soluble in hydrochloric acid and nitric acid. With a solution of caustic potassa, it blackens.

Beramite is of similar character occurring in small foliated, columnar masses. H = 2. Color, hyacinth-red to reddish-brown; streak, dirty yellow.

§ **166. Scorodite.** — Fe^2O^3, $AsO^5 + 4HO$. 34.7Fe^2O^3, 49.8AsO^5. H = 3.5—4. G = 3.1—3.3. Orthorhombic. Crystallized. Color pale leek-green or liver-brown. Lustre, vitreous; streak, white.

In a matrass, yields pure water. In the forceps, fuses to a gray scoriaceous slag of metallic lustre, coloring the outer flame pale-blue. On Ch., emits arsenical vapors and fuses to a gray magnetic slag, of metallic lustre, which gives with fluxes the reactions of iron.

Not affected by nitric acid; forms a brown solution with hydrochloric acid; partially dissolved by ammonia, leaving a brown residue.

ORES OF LEAD.

§ **167. Massicot** [Plumbic Ochre]. — PbO., containing frequently $PbO.CO^2$, CaO, Fe^2O^3, and SiO^2. 92.83Pb. G = 8. Orthorhombic, also isometric. Massive. Lustre, dull; color, between sulphur and orpiment-yellow. Streak, light-yellow.

Before the Blp., behaves like oxide of lead.

§ **168. Minium.** — $PbO^2 + 2PbO$. G = 4.6. Pulverulent. Color, vivid-red, mixed with yellow.

Before the Blp., behaves like oxide of lead.

With hydrochloric acid evolves chlorine, and is converted into chloride of lead. With nitric acid, becomes brown.

§ **169. Galenite.** — PbS. 86.6Pb. H = 2.5—2.75. G = 7.25—7.7. Isometric. Color and streak, lead-gray; of metallic lustre. Crystals usually affect the cubical form, and possess very perfect cubic cleavage. It is generally argentiferous.

Heated in a matrass, sometimes decrepitates and frequently yields a slight white sublimate. Heated in an open glass tube, emits sulphurous acid, and, on the heat being raised, gives a white sublimate of sulphate of lead. Heated on Ch., affords a globule of pure lead, the Ch. becoming at the same time covered with sulphate of lead and oxide of lead. The globule of metallic lead yields generally a little silver on cupellation. The presence of antimony is ascertained as shown § 49. Zinc, § 113. Iron, § 83.

It dissolves with some difficulty in boiling hydrochloric acid, with evolution of sulphuretted hydrogen. Very dilute nitric acid has no effect on it, but by a stronger acid it is readily dissolved with evolution of nitrous acid vapors. By fuming nitric acid and aqua regia it is very violently acted upon, being converted into sulphate, or a mixture of the sulphate with the chloride.

§ **170. Bournonite.** — $(3Cu^2S, Sb^2S^3) + 2(3PbS, Sb^2S^3)$. 24.4Pb, 25Sb, 12.9Cu. H = 2.5—3. G = 5.7—5.9. Orthorhombic. Occurs crystallized, and massive, granular, compact; lustre, metallic; color and streak, steel-gray.

In a matrass decrepitates, and yields with a strong heat a dark-red sublimate. In an open tube, sulphurous acid is evolved and abundant antimonial fumes, which condense partly on the upper and partly on the lower side of the tube; the former consists of antimonous oxide, which is volatile; the latter is not volatile, and consists of a mixture of antimonate of antimony with antimonate of lead. On Ch., fuses readily to a black globule and deposits a Ct. of antimonous oxide; with strong heat a Ct. of oxide of lead is obtained; the remaining globule, when treated with Bx. in O. Fl., gives the reactions of copper, and the globule assumes the appearance of metallic copper.

Dissolves readily in nitric acid to a blue liquid, leaving a residue of antimonous oxide and sulphur. Aqua regia leaves a residue of sulphur, chloride of lead, and antimonite of lead; the solution gives a precipitate with water. Ammonia dissolves a portion of the sulphide of antimony.

The following ores behave before the Blp. in a very similar manner:

Geocronite. — $5PbS + (Sb, As)^2S^3$. Orthorhombic.

Dufrenoysite. — $2PbS + As^2S^3$. Orthorhombic.

Boulangerite.— $3PbS + Sb^2S^3$.

Jamesonite. — $2(FePb)S + Sb^2S^3$. Orthorhombic.

Plagionite. — $PbS + Sb^2S^3 + \frac{1}{4}PbS$. Monoclinic.

Zinkenite. — $PbS + Sb^2S^3$. Orthorhombic.

Meneghenite. — $4PbS + Sb^2S^3$. Orthorhombic.

Those minerals in which a part of the Sb^2S^2 is substituted by As^2S^3, give on Ch. arsenical vapors, and in an open tube a crystalline sublimate.

§ 171. **Phosgenite** [Corneous Lead].—$PbCl + PbO.CO^2$. 73.8Pb. $H = 2.75—3$. $G = 6—6.3$. Orthorhombic. Forms crystals of adamantine lustre, of white, gray, or yellow color.

In a matrass, decrepitates slightly and becomes a little darker yellow. On Ch., fuses readily, emits acid vapors, becomes reduced to metallic lead, and gives a white Ct. of chloride of lead and a yellow Ct. of oxide.

Dissolves in nitric acid with effervescence.

§ 172. **Cerussite** [White Lead Ore]. — PbO. CO^2. 77.6Pb. H = 3 — 3.5. G = 6.4. Orthorhombic. Occurs granularly massive, or in prismatic needles, or compressed plates. Color, mostly white, yellow, or gray. Streak, colorless.

When heated in a matrass, decrepitates and turns yellow; carbonic acid is given out. Heated on Ch. alone, is reduced to metallic lead. Treated with fluxes, dissolves with effervescence and gives the reactions of pure oxide of lead (v. Table II., 15); dissolves readily and with effervescence in dilute nitric acid; with hydrochloric acid, leaves a residue of chloride of lead; dissolves in a solution of caustic potassa.

§ 173. **Leadhillite.** — PbO.SO^3 + 3(PbO.CO^2). 75.Pb. H = 2.5. G = 6.2 — 6.5. Orthorhombic. Occurs in transparent crystals of pearly or resinous lustre. Color and streak, white, passing into yellow, green, or gray.

On Ch., intumesces slightly, becomes yellow, but white again on cooling; with greater heat easily reduced to metallic lead.

Dissolves in nitric acid with effervescence, leaving a residue of sulphate of lead.

§ 174. **Anglesite** [Lead Vitriol]. — PbO.SO^3. 68.3Pb. H = 2.75 — 3. G = 6.2. Orthorhombic. It often occurs in small octahedral crystals with many facets, but more frequently in lamellar masses; of high lustre.

Heated in a matrass, decrepitates and usually yields a little water. Treated on Ch. in O. Fl., fuses to a clear bead, which on cooling turns milk-white; with Sd. on Ch., affords a globule of metallic lead; the Sd. is absorbed by the Ch. and shows, when placed on silver-foil, a strong sulphur reaction. With the fluxes, gives the reactions of oxide of lead. Traces of iron or manganese may be detected by Bx. or Sd. as shown §§ 83 and 90.

It dissolves in acids only with great difficulty; by hydrochloric acid it is partly decomposed; the pulverized mineral is soluble in a solution of caustic potassa.

§ **175. Pyromorphite** [Phosphate of Lead].—Essentially $PbCl + 3(3PbO.PO^5)$. 76.2Pb. $H = 3.5 - 4$. $G = 6.5 - 7$. Hexagonal. It occurs often in globular masses with a columnar structure, also fibrous and granular. Lustre, adamantine. Streak, slightly yellow. Color, green, yellow, and brown.

Heated in a matrass, sometimes decrepitates and yields, with continued heat, a faint white and volatile sublimate of chloride of lead. Heated in the platinum-pointed pincers, fuses readily and colors the outer flame blue; if the amount of phosphoric acid is not too small, the edges of the flame will appear green. With S. Ph. and oxide of copper, gives the reaction for chlorine, § 65. On Ch. in the O. Fl., fuses to a globule, which on cooling assumes a polyhedral form and a dark color; in the R. Fl., yields a Ct. of oxide of lead, and the globule, on cooling, assumes dodecahedral facets of pearly lustre. With boracic acid and iron wire, gives the reaction for phosphoric acid (§ 95). With Sd. on Ch., affords metallic lead. Some varieties, a portion of the phosphoric acid is replaced by arsenous oxide, which is readily detected by the odor when treated with Sd. on Ch. (§ 54). Also a part of the lead is replaced by lime, as in the brown varieties polysphærite, miesite, and musierite, while some of the Pb.Cl.

is replaced by calcium fluoride, thus diminishing the amount of lead.

Soluble in nitric acid, and solution of caustic potassa.

§ 176. **Plumbo-Gummite.** — $3(3PbO, PO^5) + 6(Al^2O^3, 3HO)$. $H = 4 - 4.5$. $G = 6.3 - 6.4$. In reniform or globular masses, with a columnar structure; also, compact massive. Of resinous lustre; color usually yellowish-brown; resembling gum-arabic in appearance. Streak, colorless.

In a matrass, decrepitates and gives out water. In the forceps, intumesces and colors the outer flame azure-blue. On Ch., intumesces, becomes white and opaque, and fuses but imperfectly, depositing a faint white Ct. of chloride of lead. In small quantities, soluble in Bx. and S. Ph. to clear beads. With Sd. on Ch., minute globules of metallic lead are obtained. Treated with So. Co., assumes a fine blue color.

Soluble in nitric acid.

§ 177. **Crocoite** [Red Lead Ore].—PbO, CrO^3. 63.2Pb. $H = 2.5 - 3$. $G = 5.9 - 6.1$. Monoclinic. Occurs usually in bright hyacinth-red crystals of adamantine lustre. Streak, orange.

In a matrass, decrepitates; the crystals are broken up into minute pieces and assume a darker color. On Ch., fuses and becomes reduced with detonation; a Ct. of oxide of lead is formed, and grayish-green sesquioxide of chromium remains with the metallic globule. With Sd. on Ch., affords a globule of metallic lead. With Sd. on platinum foil, fuses to a dark-yellow mass, which becomes green in R. Fl. With Bx. or S. Ph. in O. Fl., dissolved; the bead appears yellow while hot, but becomes green on cooling. Fused in a platinum spoon with from three to four parts of bisulphate of potassa, gives a dark-violet mass, which is greenish-white when cold.

Dissolves in heated hydrochloric acid to a green liquid, leaving a residue of chloride of lead. Dissolves with difficulty in nitric acid to a yellowish-red liquid. A solution of caustic potassa colors it brown, and finally dissolves it to a yellow liquid.

§ 178. **Vauquelinite.** — $3CuO,2CrO^3 + 2(3PbO,2CrO^3)$. 56.4Pb, 8.6Cu. H = 2.5—3. G = 5.5—5.7. Monoclinic. Occurs usually in minute crystals, or in reniform or granular masses. Color, dark-green to brown, sometimes nearly black. Lustre, adamantine to resinous; streak, greenish to brownish.

On Ch., fuses with effervescence to a gray submetallic globule; where the mass is in contact with the coal, small globules of lead make their appearance; in R. Fl. a Ct. of oxide of lead is formed. With Bx. or S. Ph. in O. Fl., clear green beads are obtained, which remain green on cooling, but which on application of the R. Fl. become red and opaque; this reaction appears most distinctly on Ch. with Sn. With Sd. on platinum wire in O. Fl., dissolves to a transparent green bead, which on cooling becomes yellow and opaque; on treating the bead with a few drops of water, a yellow solution is obtained, in which the presence of chromic acid may be proved as described § 68. With Sd. on Ch., is completely decomposed; on treating the reduced metals with boracic acid on Ch. (v. § 71) a globule of metallic copper is obtained.

Partly soluble in nitric acid to a dark-green liquid; the residue is yellow.

§ 179. **Wulfenite** [Yellow Lead Ore]. — PbO, MoO^3, sometimes with a little CrO^3. 57Pb. H = 2.75—3. G = 6.3—6.9. Tetragonal. Crystallized or granularly massive, firmly coherent. Color, usually wax-yellow, passing into orange-yellow. Streak, white.

In a matrass, decrepitates and becomes darker while hot. On Ch., fuses and is partly absorbed by the coal, while metallic lead and a Ct. of oxide of lead are deposited. With Bx. or S. Ph. on platinum wire gives the reactions of molybdic acid (v. Table II., 18). With Sd. on Ch., affords a globule of metallic lead. Fused with bisulphate of potassa in a platinum spoon, a yellowish mass is obtained, which becomes white on cooling; treated with distilled water and a piece of metallic zinc placed in the solution, the liquid assumes a blue color.

Dissolves in concentrated hydrochloric acid to a green liquid, leaving a residue of chloride of lead. The pulverized mineral is decomposed on being digested with nitric acid; a yellowish-white residue is left, which becomes blue when exposed to air in thin layers.

ORES OF MANGANESE.

§ **180. Pyrolusite** [Gray Ore of Manganese]. — MnO^2. 63.3Mn. H = 2—2.5. G = 4.8. Orthorhombic. Of black or dark-gray color and little lustre; powder black; sometimes of columnar structure. Streak, black or bluish-black.

In a matrass, usually yields a little water; when heated to redness, oxygen is evolved. Alone infusible, but turning reddish-brown when the temperature is sufficiently high. Soluble in Bx. and S. Ph. with the usual manganese reactions; gives frequently the indications of iron.

Soluble in hydrochloric acid with disengagement of chlorine.

§ **181. Hausmannite** [Black Manganese]. — $(MnO)^2MnO^2$ 72.1Mn. H = 5 — 5.5. G = 4.7. Tetragonal. Crystallized, or granular, particles strongly coherent. Color, brownish-black; streak, chestnut-brown.

Before the Blp., and with hydrochloric acid, behaves like the preceding ore.

§ 182. **Braunite.** — $2Mn^2O$, $MnO^2 + MnO^2SiO^2$. H = 6—6.5. G = 4.7—4.8. Tetragonal. Occurs crystallized or massive. Color and streak, dark brownish-black.

In a matrass, does not give any water; behaves otherwise like pyrolusite. Dissolves in hydrochloric acid with disengagement of chlorine, leaving sometimes a residue of silica.

§ 183. **Psilomelane.** — Composition very various, essentially $Mn^2O^3 + MnO^2$ with BaO, or KO, and HO. H = 5—6. G = 3.7—4.3. Massive, and botryoidal. Color, iron-black; streak, brownish-black, shining.

Before the Blp. and with solvents it behaves like pyrolusite.

§ 184. **Wad** [Bog Manganese]. — Essentially MnO^2, MnO, and HO; contains often Fe^2O^3, Al^2O^3, BaO, SiO^2, etc. H = 0.5—6. G = 3—4.2. Amorphous, earthy, or compact, of a dull black color.

In a matrass, yields water abundantly, and otherwise behaves like pyrolusite. Some varieties, known under the name of "Cupreous Manganese," when treated with Sd. and Bx. on Ch., afford a globule of metallic copper.

§ 185. **Rhodochrosite.** — MnO,CO^2 when pure, sometimes (MnO, FeO, CuO, MgO), CO^2. H = 3.5—4.5. G = 3.4—3.6. Rhombohedral. Occurs crystallized, or in globular masses of columnar structure; also massive. Color, shades of rose-red, brownish; streak, white.

In a matrass, some varieties give a little water and decrepitate violently. Infusible. Some varieties, when heated in R. Fl., become magnetic. Dissolves in fluxes with effervescence, and gives usually the reaction of manganese and iron.

The pulverized mineral is little affected by hydrochloric acid in the cold; on heating, dissolves with effervescence.

§ **186. Franklinite.**—(FeO, ZnO, MnO), (Fe^2O^3, Mn^2O^3) usually with a little SiO^2, Al^2O^3. H = 5.5—6.5. G = 5. Isometric. Occurs crystallized, and massive. Lustre, metallic; color, iron-black; streak, dark reddish-brown; acts slightly on the magnet.

Infusible. Dissolves in Bx. and S. Ph. with manganese reaction; the Bx. bead, when treated on Ch. in R. Fl., becomes bottle-green. With Sd. on platinum foil, gives manganese reaction. With Sd. on Ch., gives a faint Ct. of oxide of zinc, which becomes more distinct on addition of Bx.

Dissolves completely in heated hydrochloric acid to a greenish-yellow liquid, chlorine being evolved.

ORES OF MERCURY.

§ **187. Native Mercury.** — Hg., sometimes containing a little Ag. G = 13.5. Metallic globules of a tin-white color.

Heated in a matrass, is converted into vapor, which condenses in the neck of the matrass to small metallic globules.

Dissolves readily in nitric acid.

§ **188. Amalgam.** — $AgHg^2$, 65.2Hg, and $AgHg^3$. 73.75Hg. H = 3—3.5. G = 10.5—14. Isometric. Occurs crystallized and massive. Color and streak, silver-white; opaque.

In a matrass, boils, gives a sublimate of metallic mercury, and leaves a spongy residue of silver, which on Ch. fuses readily to a globule.

Dissolves readily in nitric acid.

Arquerite. — Ag6Hg. 13.5Hg. Isometric. In color, lustre; ductility like native silver, but softer.

§ **189. Calomel** [Horn Quicksilver]. — Hg^2Cl. 84.9Hg. H = 1—2. G = 6.48. Tetragonal. Occurs usually in distinct crystals, or crystalline coats, of adamantine lustre and yellowish-gray color. Translucent; streak, pale yellowish-white.

In a matrass, yields a white sublimate of subchloride of mercury. Mixed with Sd. and heated in a matrass, affords globules of metallic mercury. On Ch., completely volatilized, giving a white Ct. Shows the chlorine reaction when treated as described § 65.

Treated with boiling hydrochloric acid, is partly dissolved and becomes gray. Not affected by nitric acid; dissolved by aqua regia. With a solution of caustic potassa, becomes black.

§ **190. Cinnabar.** — HgS. H = 2—2.5. G = 8.9. Rhombohedral. Color, various shades of red, from cochineal-red to dark brownish-red. Powder always bright-red. It occurs in very small flattened crystals, or granularly massive. Streak, scarlet, subtransparent to opaque.

Heated in a matrass, is volatilized and condenses to a black sublimate, which by friction assumes a red color. Mixed with Sd., yields on heating globules of metallic mercury. In an open glass tube, is partially decomposed into metallic mercury and sulphurous acid. On Ch. it is, when pure, wholly volatilized.

Nitric acid and hydrochloric acid have no visible effect on it. Aqua regia dissolves it, part of the sulphur being precipitated. Insoluble in caustic potassa.

ORES OF NICKEL.

§ **191. Niccolite** [Copper Nickel]. — Ni, As, or Ni^3As^3. 44.1Ni, 55.9As. Sometimes part of the As. replaced by

Antimony. H = 5—5.5. G = 7.3—7.6. Hexagonal. Usually massive; of copper-red color, with a gray tarnish, and metallic lustre; very brittle. Streak, dark-brown.

In a matrass, affords a very slight sublimate of arsenous oxide. In an open glass tube, yields a copious sublimate of arsenous oxide, and usually a little sulphurous acid; the assay-piece assumes at the same time a yellowish-green color and crumbles to powder. On Ch., emits arsenical fumes and fuses to a white and brittle globule which, when treated with Bx. in R. Fl., imparts usually to the flux the colors of iron and cobalt. Sometimes a faint Ct. of oxide of lead is deposited on the Ch.

Dissolves almost completely in concentrated nitric acid; the solution has a green color; on cooling, arsenous oxide separates. Readily dissolved by aqua regia.

§ 192. **Gersdorffite** [Nickel Glance]. — $NiS^2 + NiAs^2$, or $Ni(S,As)^2$. 35.2Ni, 45.5As. H = 5.5. G = 5.6—6.9. Isometric; pyritohedral. Of silver-white or steel-gray color, and metallic lustre. Streak, grayish-black.

In a matrass, decrepitates violently and yields a yellowish-brown sublimate of sulphide of arsenic. In an open glass tube, emits arsenous oxide and sulphurous acid. On Ch., fuses with emission of sulphur and arsenical fumes to a globule which, when treated with Bx. in R. Fl., gives the reactions of iron and cobalt. After having removed these two metals, the remaining globule exhibits with the fluxes the reactions of pure oxide of nickel.

Partly dissolved by nitric acid, sulphur and arsenous oxide being precipitated.

§ 193. **Ullmannite** [Nickeliferous Gray Antimony]. — $NiS^2 + Ni(Sb,As)^2$. 27.6Ni. The arsenic is sometimes wanting. H = 5—5.5. G = 6.2—6.5. Isometric. It closely resembles the preceding ore in its physical properties. Streak, dark-gray.

In a matrass, yields a slight white sublimate. In an open glass tube, emits copious antimonial fumes and sulphurous acid. On Ch. in R. Fl., fuses to a globule, and coats the Ch. with antimonous oxide; sometimes the odor of arsenic is observable. The melting globule, when treated with Bx., frequently exhibits the reactions of iron and cobalt besides those of nickel.

It is violently acted upon by concentrated nitric acid, sulphur, antimonous and arsenous oxides being precipitated. Aqua regia dissolves it, excepting the sulphur, to a green liquid.

§ 194. **Millerite** [Capillary Pyrites].—NiS. 64.9Ni. H = 3—3.5. G = 5.2—5.6. Rhombohedral. Occurs usually in delicate capillary crystals of brass-yellow color and metallic lustre. Streak, bright.

In an open glass tube, evolves sulphurous acid. On Ch., fuses with emission of sparks to a metallic globule which is attracted by the magnet. The calcined mineral gives with fluxes the indications of oxide of nickel, and sometimes also those of oxide of cobalt.

By heated concentrated nitric acid it is but little affected, but its color is changed to gray. By aqua regia it is wholly dissolved.

§ 195. **Zaratite** [Emerald Nickel].—$(NiO.CO^2 + 4HO) + 2(NiO.HO)$. 46.5NiO. H = 3—3.2. G = 2.5—2.7. Usually forms incrustations of emerald-green color, and vitreous lustre. Streak, pale-green.

In a matrass, loses already at 212° a considerable amount of water, and blackens. In Bx. and S. Ph., dissolves with effervescence, exhibiting the characteristic nickel reactions.

Dissolves easily in heated dilute hydrochloric acid with effervescence.

§ 196. **Annabergite** [Nickel Green]. — $3NiO.AsO^5 + 8H.O.$ 29.2Ni. Monoclinic. Soft. In capillary crystals, also massive and disseminated. Color, fine apple-green. Streak, somewhat lighter.

In a matrass, yields water and darkens in color. In the forceps, fuses and colors the outer flame light-blue. On Ch. in R. Fl., fuses with emission of arsenical vapor to a blackish-gray globule; when treated with Bx. the globule gives the reactions of nickel, sometimes also those of iron and cobalt.

Soluble in acids.

ORES OF SILVER.

§ 197. **Native Silver.**—Pure silver, associated with gold, copper, arsenic, iron, and other metals. H = 2.5—3. G = 10—11. Isometric. Color, silver-white; lustre, metallic; ductile and malleable. Occurs usually in twisted filaments, or arborescent; sometimes in plates or massive.

On Ch., fuses easily to a globule, which assumes a bright surface, and shows after cooling a silver-white color. Foreign metals are detected by the methods given §§ 103–105.

It dissolves in nitric acid.

§ 198. **Dyscrasite** [Antimonial Silver]. — $Ag^3Sb.$ and $Ag^2Sb.$ H = 3.5—4. G = 9.4—9.8. Orthorhombic. Occurs crystallized or massive; granular. Lustre, metallic; color and streak, silver-white.

On Ch., fuses readily to a gray non-ductile globule and coats the Ch. with oxide of antimony; with continued heat the globule assumes the appearance of pure silver, and the Ct. becomes reddish.

Dissolves in nitric acid, leaving a residue of oxide of antimony.

§ 199. **Cerargyrite** [Horn Silver]. — AgCl. 75.2Ag. H = 1—1.5. G = 5.5. Isometric. Remarkable for its pearl-gray or greenish color, its semi-transparency, resinous lustre, and more especially for its softness, which is so great as to allow it to be marked by the nail. It turns brown on exposure to air. When rubbed with a moistened plate of zinc or iron the latter becomes covered with a coating of silver. The streak is shining.

It fuses in a candle-flame. On Ch., is easily reduced, especially when mixed with Sd. Mixed with oxide of copper and heated on Ch. in R. Fl., chloride of copper is formed, which colors the flame azure-blue (v. § 65).

Insoluble in water and nitric acid. Slowly soluble in caustic ammonia. Partially decomposed by a boiling solution of caustic potassa.

§ 200. **Embolite** [Chloro-Bromide of Silver]. — AgBr. and AgCl. in varying proportions. 61 to 69.8Ag. H = 1—1.5. G = 5.3—5.4. Isometric. Crystallized or massive. Lustre, resinous; color, various shades of green.

On Ch., fuses readily, evolves pungent vapors of bromine, and affords a globule of metallic silver. With Sd. on Ch., reduced; on dissolving in water the alkaline mass which has passed into the coal, evaporating the solution to dryness, and treating the residue with bisulphate of potassa as described § 63, bromine vapors are given out. Fused with oxide of copper on Ch. in R. Fl., colors the outer flame greenish, then blue (v. § 65).

§ 201. **Bromyrite** [Bromic Silver]. — AgBr. 57.4Ag. H = 2—3. G = 5.8—6. Isometric. Occurs usually in small concretions. Lustre, splendent; color, yellowish-green or green. Sectile.

11

Before the Blp. on coal emits bromine vapors and yield a globule of silver. Fused with bisulphate of potash, in a matrass, gives off yellowish-brown vapors of bromine. Insoluble in nitric acid; sufficiently soluble in ammonia.

§ 202. **Iodyrite** [Iodic Silver].—AgI. 46Ag. Soft. G = 5.5. Hexagonal. Occurs crystallized or in thin plates with a lamellar structure. Color, citron-yellow to yellowish-green. Lustre, resinous to adamantine.

On Ch., fuses readily, colors the flame purple-red, and affords a globule of silver.

§ 203. **Argentite** [Silver Glance]. — AgS. 87.1Ag. H = 2—2.5. G = 7. Isometric. Color, blackish lead-gray; lustre, metallic. It is easily distinguished from other minerals of the same color by being cut by a knife like lead.

On Ch. in O. Fl., intumesces, gives out sulphurous acid, and finally yields a globule of metallic silver.

Soluble in dilute nitric acid, leaving a residue of sulphur.

Jalpaite.—A cupriferous silver glance from Mexico. Color, blackish lead-gray.

Acanthite.—AgS. Orthorhombic. Reactions the same as for Argentite.

§ 204. **Pyrargyrite** [Dark-red Silver Ore].—3AgS, SbS^3. 59.9Ag. H = 2—2.5. G = 5.7—5.9. Rhombohedral. Color, dark-red to black, giving a cochineal-red powder. Crystallizes in hexagonal prisms. Streak, cochineal-red. Lustre, metallic-adamantine.

In a matrass, fuses very readily and yields with continued heat a sublimate of tersulphide of antimony. In an open glass tube, gives antimonial fumes and sulphurous acid. On Ch., fuses readily and deposits a Ct. of antimonous oxide, being converted into sulphide of silver; if for a long time exposed to the O. Fl., or, when mixed with Sd., in the R. Fl., affords a globule of metallic silver.

Part of the SbS^3 is sometimes substituted by AsS^3; it then gives out arsenical fumes when mixed with Sd. and heated in the R. Fl. on Ch.

The pulverized mineral, when heated with nitric acid, turns black, and is ultimately dissolved, leaving a residue of sulphur and antimonous acid. Caustic potassa also blackens it and effects partial solution, from which acids precipitate tersulphide of antimony.

§ **205. Proustite** [Light-red Silver Ore]. — $3AgS, As^2S^3$. 65.4Ag. H = 2—2.5. G = 5.4—5.5. Rhombohedral. Very much resembles the dark-red silver ore, but is of a somewhat lighter color. Lustre, adamantine.

Before the Blp. and to solvents, behaves like the preceding, excepting it gives off arsenical fumes instead of antimonous oxide. The solution in caustic potassa deposits a yellow precipitate when neutralized with acids.

§ **206. Stephanite** [Brittle Silver Ore]. — $5AgS, Sb^2S^3$. 68.5Ag. H = 2—2.5. G = 6.2. Orthorhombic. Of metallic lustre and iron-black color and streak; it is very brittle and fragile.

In a matrass, decrepitates, then fuses and ultimately yields a faint sublimate of tersulphide of antimony. On Ch., fuses very readily and coats the Ch. with antimonous acid. If the blast with the O. Fl. is kept up for a sufficient time, the Ct. assumes a red color and a globule of metallic silver is obtained. Contains frequently copper and iron, which may be detected by the process described § 71. If arsenic is present it gives in the open tube a crystalline sublimate of arsenous acid.

In dilute heated nitric acid it dissolves, excepting the sulphur and antimonous oxide; the solution becomes milky on addition of water. Partially dissolved by a boiling solution of caustic potassa.

§ **207. Polybasite.** — 9 (Ag, Cu^2)S + (Sb, $As)^2S^3$. H = 2—3. G = 6.2. Orthorhombic. Occurs usually in short tabular prisms, or massive. Lustre, metallic; color and streak, iron-black.

In a matrass, fuses very readily, but gives nothing volatile. In an open tube, gives sulphurous acid and antimonial fumes; the sublimate contains sometimes crystals of arsenous acid. On Ch., gives a Ct. of oxide of antimony; with continued heat, gives a bright metallic globule, which, on cooling, becomes black on its surface; sometimes a faint Ct. of oxide of zinc is deposited; the metallic globule affords with fluxes the reaction of silver and copper.

With acids behaves like bournonite.

§ **208. Stromeyerite** [Argentiferous Sulphide of Copper]. — Cu^2S. + AgS. 53Ag, 31.2Cu. H = 2.5—3. G = 6.2—6.3. Orthorhombic. Occurs usually in small, compact masses. Lustre, metallic; color, dark steel-gray. Streak, shining.

In a matrass, fuses easily and gives sometimes a little sulphur. In an open tube, fuses to a globule and gives sulphurous acid. On Ch., fuses to a gray metallic globule which is a little malleable; with fluxes the globule gives the reactions of copper, sometimes also those of iron; on a cupel with lead affords a globule of silver.

Dissolves in nitric acid, leaving a residue of sulphur.

ORES OF TIN.

§ **209. Cassiterite** [Tin Ore]. — SnO^2. 78.67Sn. H = 6—7. G = 6.3—7.1. Tetragonal. It occurs crystallized in square prisms terminated by more or less complicated pyramids; re-entrant angles are so frequent that they are to a certain extent characteristic; also massive, and in small mammillated masses of fibrous texture, hence called "wood tin." Color, very various, but usually brown or black. The crystals commonly possess a very brilliant lustre.

Infusible in the forceps; the behavior before the Blp. is that of pure oxide of tin (v. Table II., 30), excepting that it sometimes imparts to the Bx. bead a slight yellowish tinge, owing to the presence of iron, and exhibits the reaction for manganese when fused with soda and nitre on platinum-foil.

Insoluble in acids. Fused with caustic potassa, yields a mass which is mostly soluble in water.

§ **210. Stannite** [Tin Pyrites]. — 2 $(Cu^2, Fe, Zn)S + SnS^2$. 30.5Sn. H = 4. G = 4.3—4.5. Probably tetragonal and hemihedral. Of steel-gray or iron-black color, and metallic lustre. Occurs usually massive, granular, and disseminated. Streak, blackish.

In an open glass tube, yields sulphurous acid and oxide of tin, which collects close to the assay-piece, and which cannot be volatilized by heat. On Ch. in R. Fl., fuses to a black scoriaceous globule; in O. Fl., gives out sulphurous acid and becomes covered with oxide of tin. When well calcined by the alternate application of O. Fl. and R. Fl., gives with Bx. the indications of Fe. and Cu. With Sd. and Bx., yields a globule of impure copper.

Decomposed by nitric acid, a blue solution is obtained, and a mixture of sulphur and oxide of tin remains undissolved.

ORES OF ZINC.

§ **211. Zincite** [Red Zinc Ore]. — ZnO, containing some Mn^2O^3, or MnO^2. 80.26Zn. H = 4—4.5. G = 5.4—5.5. Hexagonal. Of a deep-red color and high lustre; of distinctly foliated structure, and orange-yellow streak.

Infusible alone. Dissolved by Bx. in O. Fl. with manganese reaction. With Sd. on Ch., deposits a copious Ct. of oxide of zinc.

Soluble in nitric acid without effervescence; in hydrochloric acid with evolution of chlorine.

§ **212. Sphalerite** [Blende]. — ZnS. 67.Zn. H = 3.5—4. G = 3.9—4.2. Isometric. Of very variable color, from yellow to black; of resinous lustre and lamellar aspect, distinctly cleavable. It occurs often crystallized in rhomboidal dodecahedrons. The powder is always light-colored, white or grayish, and dull.

In a matrass, sometimes decrepitates violently, but gives nothing volatile; its color also remains unchanged, excepting the green varieties, which become yellow. Strongly heated in an open glass tube, sulphurous acid is evolved, and the color of the calcined assay is white, yellowish, or brownish, according to the amount of FeS. which it contains. Alone, infusible or only rounded at the thinnest edges. On Ch. in R. Fl. a feeble dark Ct. of oxide of cadmium is usually obtained, which is soon followed by a pure zinc-Ct. With Sd. on Ch., is easily reduced, and the characteristic zinc-flame may frequently be observed. Iron is readily detected by calcining the mineral in the O. Fl. and treating the residue with Bx.

The pulverized mineral dissolves in nitric acid, leaving a residue of sulphur.

§ **213. Smithsonite** [Carbonate of Zinc]. — $ZnO.CO^2$. 52Zn. H = 5. G = 4—4.5. Rhombohedral. Of vitreous lustre, and white, grayish, or brownish color and streak; semi-transparent or opaque. Often stalactitic or mammillary.

Heated in a matrass, loses carbonic acid, and, if pure, appears after cooling enamel-white. The ZnO. is often to a large extent substituted by FeO., MnO., CdO., PbO., MgO., CaO.; it then, after cooling, frequently assumes a dark color and gives with fluxes the indications of iron and manganese. Mixed with Sd. and exposed to the R. Fl., it is decomposed, and oxide of zinc deposited on the Ch. If the temperature was raised sufficiently high, a zinc-flame is sometimes observable. The Ct. is at first dark-yellow, or reddish, when cadmium is present.

It readily dissolves in acids with effervescence; also in caustic potassa.

§ **214. Calamine** [Hydrous Silicate of Zinc]. — 2ZnO, SiO^2 + HO. 67.5ZnO., with sometimes a little lead. H = 4.5—5. G = 3.1—3.9. Orthorhombic. It closely resembles in its physical characters the preceding ore. It is electric by heat; the smallest fragment heated attracts light substances.

Infusible in the forceps. In a matrass yields water and turns milk-white. Bx. dissolves it to a transparent glass, which cannot be made opaque by flaming. It dissolves in S. Ph. to a transparent glass which becomes opaque on cooling, and in which, when highly saturated, clouds of silica are observable while hot. With Sd. on Ch., swells and affords with difficulty a Ct. of oxide of zinc. With SoCo., assumes a green color, which, when the heat is raised, passes into a fine light-blue on the fused edges.

It is readily decomposed by acids, with separation of gelatinous silicic acid. Partly dissolved by caustic potassa.

§ **215. Willemite** [Anhydrous Silicate of Zinc].—2ZnO, SiO^2, and often containing a little Mn^2O^3, Fe^2O^3, CaO, and MgO. 72.9Zn. H = 5.5. G = 3.89—4.18. Rhombohedral. Lustre, vitreo-resinous; weak. Color, whitish or greenish-yellow, when purest; green to dark-brown when impure. Streak, uncolored. Transparent to opaque. Brittle.

Before the Blp., in the forceps glows and fuses with difficulty to a white enamel; the varieties from New Jersey fuse from 3.5 to 4. The powdered mineral, on coal in the R. Fl., gives a coating, yellow while hot, and white on cooling, which, moistened with So.Co., and treated in O. Fl., is colored bright-green. With soda the coating is more easily obtained.

Decomposed by hydrochloric acid with separation of gelatinous silica.

FOSSIL FUEL, AND CARBONACEOUS COMPOUNDS.

§ 216. **Graphite** [Plumbago]. — C. Hexagonal. In flat six-sided tables. H = 1 — 2. G = 2.089. Lustre, metallic. Streak, black and shining. Color, iron-black to dark steel-gray. Opaque. Sectile; marks paper. Thin; laminæ flexible. Feel, greasy.

It occurs also foliated, columnar, radiated, scaly, granular and massive.

At a high temperature it burns without flame or smoke, leaving usually some red oxide of iron. Before the blowpipe infusible; fused with nitre in a platinum spoon deflagrates, converting the reagent into carbonate of potash, which effervesces with acids. Unaffected by acids.

§ 217. **Anthracite.** — C., from 80 to 95 per cent., with a small percentage of SiO^2, Al^2O^3, and Fe^2O^3. H = 2—2.5. G = 1.3—1.8. Lustre, bright, often sub-metallic; color, iron-black, frequently iridescent. Fracture, conchoidal.

In a matrass, gives usually a little water, but no empyreumatic oil. Heated on platinum foil in O. Fl., is slowly consumed without flame, leaving a small quantity of ash, which consists of SiO^2, Al^2O^3, and more or less Fe^2O^3. Does not color a boiling solution of caustic potassa.

§ 218. **Bituminous Coal.** — C., H., O., in variable proportions; the bituminous matter contains from 76 to 90 per cent. of carbon; the earthy impurities consist principally of SiO^2, Al^2O^3, and CaO; contains frequently a small amount of N. and FeS^2. Softer than anthracite, G = 1.2—1.5. Less highly lustrous than the preceding, and of a more purely black or brownish-black color.

In a matrass, some varieties soften and cake (*caking coal*), while others are entirely infusible; all varieties are decomposed, evolve combustible gases and empyreumatic oils, and leave a residue of more or less metallic lustre (coke), which behaves like anthracite. On platinum foil, burns with a luminous flame and emission of smoke, leaving an earthy residue.

Boiled with a solution of caustic potassa, or with ether, imparts to these solvents no, or only a pale-yellow, color.

§ 219. **Brown Coal.** — Composition the same as that of bituminous coal, but the organic constituents contain only from 60 to 75 per cent. of carbon. In physical proportion bears sometimes a close resemblance to the preceding; some varieties show distinctly the texture of wood (*lignite*).

In a matrass, infusible, but some varieties soften; evolves combustible gases, empyreumatic oils, water of acid reaction, and a peculiar disagreeable odor, leaving a residue which consists of carbon and a considerable amount of ash. On platinum foil, burns with a smoky flame and emission of a peculiar odor.

Boiled with a solution of caustic potassa, colors the liquid brown.

§ **220. Asphaltum.** — C., H., O., in variable proportions, with about 75 per cent. of carbon. G = 1—1.2. Of black or brownish-black color, and bituminous odor.

Fuses at about 100° C., and burns with a bright flame and emission of a thick smoke, leaving little ash, which consists essentially of SiO^2, Al^2O^3, and Fe^2O^3. In a matrass, gives empyreumatic oil, some ammoniacal water, combustible gases, and leaves a carbonaceous residue.

Treated with boiling ether, colors the solvent wine-red to brownish-red (distinction from bituminous coal); treated with a boiling solution of caustic potassa, does not color the liquid, or imparts at the most a pale-yellow color (distinction from brown-coal).

§ **221. Succinite** [Amber]. — $C^{10}H^8O$ (Schrötter). H = 2—2.5. G = 1.065—1.081. It occurs in irregular masses, without cleavage. Lustre, resinous. Color, yellow; sometimes reddish, brownish, and whitish; often clouded. Streak, white. Transparent, translucent. Tasteless. Electric on friction. Fuses at 287° C., but without becoming a flowing liquid.

It consists of succinic acid, resins, an ethereal oil, and succinite proper, an insoluble substance.

Amber fuses with some difficulty in the matrass, yielding water, empyreumatic oil, gases, succinic acid, and a residue of amber-resin. It burns with a yellow flame, emitting an agreeable odor, and leaving a black carbonaceous mass.

§ **222. Elaterite** [Elastic Bitumen]. — CH^2 (mostly). G = 0.905 — 1.233. Soft, elastic, like India-rubber, but sometimes hard and brittle. Color, dark-brown. Subtranslucent.

Burns in the flame of a candle, and gives empyreumatic products when fused in a matrass.

CHAPTER V.

Systematic Method for the Determination of Inorganic Compounds.

THE careful observer, having become well acquainted with the reactions which are exhibited by the metallic oxides and other simple compounds, when subjected to the various treatments detailed in the second chapter, will find no difficulty in ascertaining the nature of any mineral substance presented to him for analysis.

If the reactions are not quite distinct, owing to an intermixture with other substances, he may call to his aid the processes laid down in the third chapter, which will enable him, in most cases, to detect also the nature of the impurities. But in order to attain satisfactory results in this way, a certain familiarity with all the principal tests is a necessary condition; this once acquired, any further directions are quite superfluous.

Those, however, who have not devoted much time to blow-pipe operations, will sometimes experience some difficulties in drawing the correct conclusions from the observed phenomena, a difficulty which is to a great extent obviated by pursuing the course given below. This methodical course has the advantage of giving the operator the answer to every phenomenon which he observes, and thus leading him, though sometimes by a very tortuous path, to the right solution. An example will show this more clearly, and teach at the same time the use of the table.

Suppose a substance be given for analysis. The operator commences with No. 1. The substance is heated in R. Fl. on Ch: a garlic odor is disengaged; proceed to No. 2. Treated with Sd. on Ch. does *not* give a mass which exhibits the reaction of sulphur; proceed to No. 3. The substance shows no metallic aspect; proceed to No. 131, thence to No. 135. It is not wholly volatilized, nor does it exhibit the reaction of sulphur; proceed to No. 137. Here we find that the substance must either be an arsenite or an arsenate (which of the two cannot be decided by the Blp. alone), and to find the metal, we proceed to No. 102. It affords, after calcination, with Sd. on Ch., a fusible metallic button; proceed to No. 103. The button is oxidizable (because on being heated in O. Fl., becomes covered with a black coating of oxide); proceed to No. 105. The button is red and malleable; the metal is copper. The substance, therefore, was arsenite, or arsenate, of copper.

The chief constituents of the body having thus been ascertained, the analyst should never omit to test the correctness of the result by the processes laid down in the third chapter. In the example given above, he should verify the result by the test given in § 57 for arsenous acid, and by those given in §§ 71 and 74 for copper. If we wish to examine the assay also for the presence or absence of some accessory constituents, we must always have recourse to the methods detailed in Chapter III. For example, having found the body under trial to consist essentially of sulphur and lead, and it appears desirable to know whether or not it contains any silver, we must subject it to the treatment described § 103.

1 { On Ch. (R. Fl.), with or without Sd., disengages a garlic odor 2
Not 4

2 { With Sd. (R. Fl.) on Ch., yields a scoriaceous mass which exhibits the sulphur-reaction (§ 107)........ *Sulpharsenide.* 131
Not 3

3 { Metallic aspect .... *Arsenide.* 131
Not........ *Arsenite and Arsenate.* 131

4 { On Ch. (O. Fl.), disengages sulphurous acid, and exhibits the sulphur-reaction (§ 107)........ *Sulphur Compound* 125
Not 5

5 { On Ch., disengages the odor of rotten horse-radish........ .... *Selenium Compound.* 136
Not........ 6

6 { The substance, after having been well dried, fuses on red-hot Ch........ 7
Not (unless strongly heated)........ 11

7 { Treated as indicated § 65, imparts to the flame an azure-blue or green color........ 8
Not........ *Nitrate.* 102

8 { The color is azure-blue........ 9
The color is green........ 10

9 { Treated as indicated § 63, disengages deep-yellow vapors. .... *Bromate.* 102
Not........ *Chlorate.* 102

10 { Treated as indicated § 63, disengages deep-yellow vapors. .... *Bromate.* 102
Violet vapors........ *Iodate.* 102

11 { Treated as indicated § 65, imparts to the flame an azure-blue or green color........ 12
Not 17

12 { Heated in a matrass with bisulphate of potassa and a little binoxide of manganese, disengages violet vapors........ .... *Iodide and Iodate.* 102
Deep-yellow vapors........ *Bromide and Bromate.* 102
Not 13

13 { Treated as indicated § 77, exhibits the fluorine-reaction........ .... *Fluoride.* 102
Not 14

14	Treated as indicated § 65, imparts to the flame an azure-blue color	15
	A green color	16
15	Heated with Sd. on Ch., gives a mass which, when mixed with bisulphate of potassa and black oxide of manganese, and heated in a closed tube, evolves a deep-yellow gas. .... *Bromide.*	102
	Not .... *Chloride and Chlorate.*	102
16	Heated with Sd. on Ch., gives a mass which, when mixed with bisulphate of potassa and peroxide of manganese, and heated in a closed tube, evolves violet vapors...... *Iodide.*	102
	Deep-yellow vapors .... *Bromide.*	102
17	Effervesces with hydrochloric acid .... *Carbonate.*	102
	Not	18
18	When finely powdered and heated with hydrochloric acid, effervesces .... *Carbonate.*	102
	Not	19
19	When finely powdered and heated with concentrated hydrochloric acid, gelatinizes	30
	Not	20
20	Fused with Sd. on Ch., yields neither a metallic globule nor a Ct.	21
	Yields a metallic globule or a Ct	22
21	Treated as indicated § 61, colors the flame yellowish-green. .... *Borate.*	102
	Not	23
22	The scoriaceous mass is heated in a platinum spoon or on platinum foil, with a drop of concentrated sulphuric acid, then alcohol poured on it, and lighted. The flame appears yellowish-green .... *Borate.*	102
	Not	23
23	Treated as indicated § 77, exhibits the fluorine-reaction...... .... *Fluoride.*	102
	Not	24
24	Heated with Sd. on Ch. yields a metallic button	25
	Not	27
25	Heated on Ch. alone behaves as indicated § 96; yields with Sd. on Ch. a soft globule .... *Phosphate of Lead.*	
	Not	26
26	The scoriaceous mass treated with boracic acid as indicated § 95, exhibits the reaction of phosphoric acid, *Phosphate.*	102
	Not	28

27 Treated as indicated § 95, exhibits the reaction of phosphoric acid..*Phosphate.* 102
Not.. 28

28 With Sd. on Ch., yields a metallic button or a copious Ct... 31
Not.. 29

29 Pulverized and fused with five or six times its weight of Sd. in a platinum spoon, or on platinum foil, yields a mass which, when heated with hydrochloric acid, gives a gelatinous precipitate.. 30
Not.. 31

30 The gelatinous precipitate placed, while still moist, on a blade of iron or zinc, becomes blue...............*Tungstate.* 139
Not.. 140

31 Metallic aspect... 32
Not.. 57

32 Yields, on Ch., a malleable and fusible metallic button, which is not oxidizable.. 33
Not.. 36

33 Yellow button.. 34
White button... 35

34 With Bx. on platinum wire, gives a bluish glass..............
....*Gold with Copper.*
Not..*Gold.*

35 With Bx. on platinum wire, gives a bluish glass..............
....*Silver with Copper.*
Not..*Silver.*

36 With Bx. on platinum wire, gives a glass which is blue in both flames..*Cobalt.*
Not.. 37

37 Gives with Bx. the reactions of oxide of copper............... 38
Not.. 42

38 Red and malleable metallic button.......................*Copper.*
Not.. 39

39 Deposits on Ch. a Ct., yellow while hot, white when cold... 41
Not; a yellowish, brittle alloy................*Copper and Tin.*

41 Malleable, yellow or reddish alloy..........*Copper and Zinc.*
White, malleable alloy.....................*Copper, Zinc, Nickel.*

42 Very fusible metallic button.. 43

43 Deposits on Ch. a Ct.. 44
No Ct. deposited on Ch.; exhibits the reactions of tin...*Tin.*

44	White Ct., very volatile............	45
	Not............	46
45	Yields on Ch. a brittle globule, which exhibits the antimony reactions. See § 30............ *Antimony.*	
	Not............ *Tellurium.*	
46	Metallic aspect, or powder assuming metallic lustre under the polishing steel............	47
	Not............	50
47	Infusible and not oxidizable............ *Platinum.*	
	Oxidizable............	48
48	With Bx. in O. Fl., an amethyst-colored glass... *Manganese.*	
	Not............	49
49	After having been oxidized, exhibits with fluxes the iron-reactions............ *Iron.*	
	After having been oxidized, exhibits with fluxes the nickel-reactions............ *Nickel.*	
50	Yields with Sd. on Ch. in R. Fl. a tin globule.. *Oxide of Tin.*	
	Not............	51
51	With Bx. on platinum wire, a green glass in both flames, ... *Chromic Iron.*	
	Not............	52
52	Yields with Sd. on platinum foil in O. Fl., a bluish-green mass............	53
	Not............	54
53	Gives with Bx. or S. Ph. on platinum wire, an amethyst-colored bead............ *Oxide of Manganese.*	
	Not. Brown powder..... *Tungstate of Iron and Manganese.*	
54	With S. Ph. on platinum wire in R. Fl., gives a glass which, on cooling, becomes brownish-red and, when touched with tin, violet-red............ *Titaniferous Iron.*	
	Not; exhibits the iron-reactions............	55
55	Heated in a closed glass tube, yields water; powder, yellow .... *Hydrate of Sesquioxide of Iron.*	
	Yields no water............	56
56	Magnetic; powder, black............ *Magnetic Iron.*	
	Not; powder, red............ *Peroxide of Iron.*	
57	Affords, with Sd. on Ch. in R. Fl., a fusible metallic button	58
	Not............	66
58	The button is malleable and not oxidizable............	59
	Oxidizable button............	60

59 { Yellow button .... *Oxide of Gold.*
White button .... *Oxide of Silver.*

60 { Button with Ct. 61
Malleable button without Ct. 75

61 { The Ct. is white and very volatile .... *Oxide of Antimony.*
Not. 62

62 { The Ct. is yellow, and the button soft 63
The button is brittle 65

63 { A very small quantity affords with Bx. or S. Ph. in O. Fl. a green glass .... *Chromate of Lead.*
Not 64

64 { The substance is yellow or reddish .... *Protoxide of Lead.*
The substance is red .... *Minium.*
The substance is brown .... *Binoxide of Lead.*

65 { Affords with Bx., or S. Ph. in O. Fl., a green bead .... *Chromate of Bismuth.*
Not .... *Oxide of Bismuth.*

66 { Treated on Ch. in O. Fl. deposits a Ct., or vaporizes completely 67
Not 70

67 { The Ct. is white, and very volatile. See § 30 .... *Oxide of Antimony.*
Not 68

68 { The Ct. is brown .... *Oxide of Cadmium.*
Not 69

69 { The substance is red or yellow, and affords, when heated in a closed glass tube, metallic mercury ... *Oxide of Mercury.*
The substance is white, becomes yellow on heating, and on cooling, white again .... *Oxide of Zinc.*

70 { Affords with Bx. a bead which is blue in both flames .... *Oxide of Cobalt.*
Not 71

71 { The Bx. bead is green in both flames 72
Not 77

72 { Soluble in water 73
Insoluble in water 74

73 { The substance is orange-red .... *Bichromate of Potassa.*
The substance is yellow .... *Chromate of Potassa or Soda.*

74 The substance is of sub-metallic aspect or grayish-black..... ....*Chromic Iron.*
The substance is a green powder...*Sesquioxide of Chromium.*

75 The button is white................................*Oxide of Tin.*
The button is red.. 76

76 The substance is red or brown.............*Suboxide of Copper.*
The substance is black...................... *Protoxide of Copper.*

77 The bead is green in O. Fl., and becomes reddish-brown in R. Fl.. 76
Not.. 78

78 The bead is amethyst-colored in O. Fl.......................... 79
Not............................ 80

79 Gives off water when heated in a glass tube.................... ....*Hydrated Oxide of Manganese.*
Not.. *Oxide of Manganese.*

80 Heated alone on Ch. in O. Fl. becomes magnetic............ 55
Not.. 81

81 Exhibits with S. Ph., on platinum wire, the uranium reactions..*Oxide of Uranium.*
Not.. 82

82 Soluble in water, exhibiting alkaline reaction................. 83
Not 91

83 Very soluble... 84
But little soluble.. 88

84 Heated on platinum wire, fuses readily and vaporizes........ 85
Not 87

85 Heated on platinum foil, stains it dark-yellow.........*Lithia.*
Not 86

86 Heated on platinum wire, colors the flame pale-violet....... ....*Hydrate of Potassa.*
Reddish-yellow; the outer flame becomes enlarged.......... ....*Hydrate of Soda.*

87 Moistened with a drop of hydrochloric acid, and heated on platinum wire, colors the flame pale-green.........*Baryta.*
Crimson..*Strontia.*

88 Moistened with a drop of hydrochloric acid, and heated on platinum wire, colors the flame crimson...........*Strontia.*
Not.. 89

89 Heated with SoCo., assumes a flesh color..........*Magnesia.*
Not.. 90

90 Heated alone, becomes very luminous.................... *Lime.*
Not; colors the flame pale-green............................ *Baryta.*

91 Heated with SoCo., assumes a fine blue color...... *Alumina.*
Not.. 92

92 Heated with SoCo., assumes a flesh color.......... *Magnesia.*
Not.. 93

93 Heated with SoCo., assumes a green color.... *Oxide of Zinc.*
Not.. 94

94 Affords with S. Ph. in O. Fl. a colorless glass, which in R. Fl. becomes blue.. 95
Not.. 97

95 Heated in a closed glass tube, evolves ammonia and becomes blue or green........................... *Tungstate of Ammonia.*
Not.. 96

96 On Ch. alone, infusible............................ *Tungstic Acid.*
Fusible.............................. *Tungstate of Potassa or Soda.*

97 Exhibits with S. Ph. the reactions of........... *Molybdic Acid.*
Not.. 98

98 Exhibits with S. Ph. the reactions of pure...... *Titanic Acid.*
Not.. 99

99 Affords with S. Ph. in R. Fl. a reddish-yellow glass; the intensity of the color increases on cooling..................... 100
Not.. 101

100 The glass, when heated on Ch. with tin, becomes violet... .... *Titanic Acid, containing Iron.*

101 With Sd. on Ch. in R. Fl., affords a metallic powder attracted by the magnet......................... *Oxide of Nickel.*
Not: affords with S. Ph. in O. Fl. a glass which, while hot, is red, and colorless when cold.......... *Oxide of Cerium.*

Nitrates, Chlorates, Bromates, Iodates, Carbonates, Phosphates, Borates, Chlorides, Bromides, Iodides, Oxides, Hydrates.

102 Affords with Sd. on Ch. in R. Fl., a fusible metallic button. 103
Not.. 109

103 The button is malleable and not oxidizable..................... 104
The button is oxidizable.. 105

104 { The button is yellow.............................. *Salt of Gold.*
The button is white.............................. *Salt of Silver.*

105 { The button is red and malleable.............. *Salt of Copper.*
Not.. 106

106 { The button is white and malleable, and forms only a faint Ct.. *Salt of Tin.*
Not.. 107

107 { Forms a white and very volatile Ct....... *Salt of Antimony.*
The Ct. is orange-yellow.. 108

108 { The button is malleable.............................. *Salt of Lead.*
The button is brittle.............................. *Salt of Bismuth.*

109 { Treated with Sd. on Ch. in R. Fl., deposits a Ct............ 110
Not.. 112

110 { The Ct. is white and very volatile. See § 30.. *Salt of Antimony.*
Not.. 111

111 { The Ct. is reddish-brown.................... *Salt of Cadmium.*
The Ct. is yellow while hot, and white when cold.......... .... *Salt of Zinc.*

112 { On Ch. alone, affords a gray and infusible powder which, under the polishing steel, assumes metallic lustre.......... .... *Salt of Platinum.*
Not.. 113

113 { Heated with Sd. in a closed glass tube, affords a sublimate of mercury.................................. *Salt of Mercury.*
Not.. 114

114 { Heated with Sd. in a closed glass tube, disengages ammonia... *Salt of Ammonia.*
Not.. 115

115 { Gives with Bx. or S. Ph. beads which are blue in both flames.. *Salt of Cobalt.*
Not.. 116

116 { The beads are green in both flames...... *Salt of Chromium.*
Not.. 117

117 { The bead exhibits the reactions produced by oxide of copper.. *Salt of Copper.*
Not.. 118

118 { Affords with Sd. on Ch. a metallic powder, which assumes lustre by friction, and is attracted by the magnet......... 119
Not.. 120

119 { Gives with Bx. in R. Fl. a bottle-green glass... *Salt of Iron.*
Gives with Bx. in R. Fl. a grayish glass...... *Salt of Nickel.*

120 { Gives with Bx. in O. Fl. an amethyst-colored bead........
.... *Salt of Manganese.*
Not.. 121

121 { Infusible mass, which assumes, with SoCo., heated in O.Fl., a fine blue color...................... *Salt of Alumina.*
A flesh color.................................. *Salt of Magnesia.*
Not.. 122

122 { The watery solution gives a precipitate on addition of some Sd.. 123
Not.. 124

123 { Heated on platinum wire, colors the flame pale-green......
.... *Salt of Baryta.*
Colors the flame crimson..................... *Salt of Strontia.*
Not; but becomes very luminous............... *Salt of Lime.*

124 { Heated on platinum wire, colors the flame violet..........
.... *Salt of Potassa.*
Colors the flame reddish-yellow................. *Salt of Soda.*

Sulphur Compounds.

125 { Metallic aspect; sulphides.................................... 126
Not.. 127

126 { The substance is calcined, and the metal detected by proceeding as indicated above, beginning with No............ 102

127 { Sulphates, hyposulphates, sulphites, hyposulphites, sulphides prepared artificially by precipitation, and a few native sulphides.. 128

128 { Heated with hydrochloric acid, disengages:
Sulphuretted hydrogen.............................. *Sulphide.* 130
Sulphurous acid 129
Nothing.......................... *Sulphate or Hyposulphate.* 130

129 { Hydrochloric acid produces a white precipitate of sulphur,
.... *Hyposulphite.* 130
Not.. *Sulphite.* 130

130 { The metal is detected, beginning with No.................... 102

Arsenic Compounds.

131 { Metallic aspect.. 132
Not.. 135

132 { Readily and completely volatilized on Ch.................... 133
Not.. 134

133 { Gives a white and very volatile Ct..............................
......................................*Arsenide of Antimony.*
Not..*Arsenic.*

134 { The substance, which is an arsenide or sulpharsenide, is thoroughly calcined, and then the metal detected as indicated above, beginning with No........................... 102

135 { Wholly volatilized on Ch., and exhibiting the reactions of sulphur.. 136
Not wholly volatilized, or exhibiting no sulphur-reaction, 137

136 { The substance is yellow..............................*Orpiment.*
The substance is red....................................*Realgar.*

137 { The substance is very volatile.................*Arsenous Oxide.*
Not: *arsenite* or *arsenate.* The substance is well calcined with alternating O. Fl. and R. Fl., and the metal found, beginning with No.. 102

Selenium Compounds.

138 { Metallic aspect...*Selenide.*
Not: *selenite* or *selenate;* the substance is well calcined, and the metal detected, beginning with No................ 102

Tungstates.

139 { With Sd. on platinum wire in O. Fl. affords a greenish-blue mass.. *Wolfram.*
Not.. 140

140 { Heated with Sd. in a closed glass tube, evolves ammonia ....*Tungstate of Ammonia.*
Not.. 141

141 { Soluble..........................*Tungstate of Potassa* or *Soda.*
Insoluble......................*Tungstate of Lime, Baryta, etc.*

Silicates.

142 { The analogy in chemical composition and properties, and the number of native silicates, make it impossible to discriminate them by a few simple tests.* The base or bases may, however, in many cases be detected by proceeding as indicated above, beginning with No.......... 102

* For the discrimination of the native silicates, v. Chapter VI.

The following scheme for the determination of inorganic substances, by Prof. T. Egleston, of Columbia College School of Mines, appeared in the American Chemist for February, 1872. In the original, the numbers referred to Prof. H. B. Cornwall's excellent translation of Plattner's Manual.

The substance may contain ***As — Sb — S — Se — Fe — Mn — Cu — Co — Ni — Pb — Bi — Ag — Au — Hg — Zn — Cd — Sn — Cl — Br — I — CO^2 — SiO^2 — NO^5 — HO,*** etc.

1. Treat on Ch. in the O. F. to find volatile substances such as ***As — Sb — S — Se — Pb — Bi — Cd,*** etc., § 20 *et seq.*

a. If there are volatile substances present, form a coating and test it with S. Ph. and tin on Ch. for ***Sb,*** § 50, or to distinguish between ***Pb*** and ***Bi,*** §§ 49 and 59.	*b.* If there are no volatile substances present, divide a part of the substance into three portions and proceed as in ***A.***

a Yellow coat, yielding, with S. Ph., a black bead; disappearing with blue flame, no part of it yielding green Sb. flame; ***Pb*** and ***Bi.***

b Yellow coat, generally with white border, yielding black or gray bead with S. Ph., disappearing with blue flame; also the border disappearing with green flame; ***Pb*** and ***Sb.***

c Yellow coat, very similar to *b*, but yielding no blue flame; ***Bi*** and ***Sb.***

2. If ***As — Sb — S — Se*** are present, roast a large quantity thoroughly on Ch. until no odor of arsenic or sulphurous acid is given off. Divide the substance into three portions, and proceed as in ***A.***

A. TREATMENT OF THE FIRST PORTION. — Dissolve a very small quantity in borax on platinum wire in the O. F. and observe the color produced. Various colors will be formed by the combination of the oxides. Saturate the bead and shake it off into the porcelain dish, repeat this once or twice. § 37 and Table III.

a. Treat these beads on Ch. with a small piece of lead, silver, or gold, in a strong R. F. §§ 71, 72, 73.

b. ***Fe — Mn — Co,*** etc., remain in the bead. § 71. If the bead spreads out on the Ch., it must be collected to a globule by continued blowing. Make a borax bead on platinum wire and dissolve in it *some* of the fragments of the bead, reserving the rest for accidents.	*c.* ***Ni — Cu — Ag — Au — Sn — Pb — Bi*** are reduced and collected by the lead button. § 71. (Sn, Pb, Bi, if present, partly volatilize.) Remove the lead button from the bead while hot, or by breaking the latter, when cold, on the anvil between paper, carefully preserving all the fragments.

d. If **Co** is present, the bead will be blue.

If a large amount of **Fe** is present, add a little borax to prove the presence or absence of **Co.** § 69.

If **Mn** is present, the bead, when treated on platinum wire in the O. F., will become dark violet or black.

e. If only **Fe** and **Mn** with no **Co** are present, the bead will be almost colorless.

Look here for **Cr, Ti, Mo, U, W, V, Ta,** by the wet way.

f. Treat the button *c* on Ch. in the O. F. until all the lead, etc., is driven off; **Ni, Cu, Ag, Au** remaining behind; or separate the lead with boracic acid. § 71.

g. Treat the residue *g* on Ch. in O. F. with S. Ph. bead, removing the button while the bead is hot.

h. If **Ni** and **Cu** are present, the bead will be green when cold. § 72. If **Ni** only — yellow. If **Cu** only — blue.

Prove **Cu** by treating with tin on Ch. in the R. F. § 71.

i. For **Ag** and **Au** make the special test.

B. Treatment of the Second Portion. — Drive off the volatile substances in the O. F. on Ch. Treat with the R. F., or mix with soda, and then treat with the R. F., for **Zn, Cd, Sn.** If a white coating is formed, test with cobalt solution. § 45.

C. Treatment of the Third Portion. — Dissolve some of the substance in S. Ph. on platinum wire in O. F., observing whether SiO^2 is present or not, and test for **Mn** with nitrate of potassa. § 90.

3. Test for **As** with soda on Ch. in the R. F., or with *dry* soda in a closed tube. § 54 *et seq.*

4. Dissolve in S. Ph. on platinum wire in the O. F. (if the substance is not metallic and does not contain any **S**), and test for **Sb** on Ch. with tin in the R. F. § 50.

5. Test for **Se** on Ch. § 99.

6. In absence of **Se** fuse with soda in the R. F. and test for **S** on silver foil. § 107. In presence of **Se** test for **S** in open tube. § 107. (To distinguish between S and SO^3, see § 108.)

7. Test for **Hg** with *dry* soda in a closed tube. Tab. II., 17.

8. Mix some of the substance with assay lead and borax glass and fuse on Ch. in the R. F. Cupel the lead button for ***Ag.*** § 105. Test with nitric acid for ***Au.*** § 79.

9. Test for ***Cl, Br,*** and ***I,*** with a bead of S. Ph. saturated with oxide of copper. §§ 62, 63, 65.

10. Test for ***Cl*** or ***Br*** with bisulphate of potassa. § 63.

11. Test for ***HO*** in a closed tube. § 9.

12. Test on platinum wire, or in platinum-pointed forceps, for coloration of the flame.

13. Test for ***C̈*** with hydrochloric acid.

14. Test for $\boldsymbol{NO^5}$ with bisulphate of potassa. § 93.

15. Test for ***Te*** in an open tube. § 109.

CHAPTER VI.

On the Determination of Minerals by means of the Blowpipe, aided by Humid Analysis.

By the methods given in the preceding chapters, we can readily detect the constituents of most inorganic compounds, whether prepared artificially or occurring in nature; especially if heavy metals form the principal constituents. But these methods do not enable us to discriminate the different native silicates, and other mineral bodies, which consist essentially of such substances as do not show any very characteristic reactions before the blow-pipe, as ex. gr. the alkaline earths. In some cases we may succeed in ascertaining the principal ingredients of the substance under examination, but fail in establishing the mineral species. To attain this end more securely, we must pursue a course, composed of an examination of the physical properties of the body and of blow-pipe operations, aided by humid analysis. The course adopted in this "Manual" is that given by *Franz von Kobell*, as laid down in his "*Tafeln zur Bestimmung der Mineralien.*"

The minerals, according to Von Kobell's system, are arranged in two large groups, the first embracing those possessing metallic lustre, the second those devoid of metallic lustre. To avoid mistakes, originating in the fact that some minerals occur sometimes with, and sometimes without, metallic lustre, these minerals will generally be found enumerated in both groups.

The same precaution has been taken in regard to those species in which the degree of fusibility, whether below or above 5, might appear doubtful. The degree of fusibility is to be judged of from the following scale:

1. **Gray Antimony.**—Fusible in coarse splinters in the flame of a candle.

2. **Natrolite.**—Fusible in fine splinters in the flame of a candle.

3. **Almandine or Iron Garnet.**—Easily fusible before the Blp.

4. **Actinolite** (a variety of hornblende).—Fusible before the Blp. in coarse splinters.

5. **Orthoclase.**—Fusible before the Blp. in fine splinters.

6. **Bronzite.**—Fusible on the edges in very fine splinters.

The fusibility, when equal to that of actinolite, is designated by 4; when between that of natrolite and almandine, by 2.5, and so on. See Appendix, page 296.

A list of the oxidized minerals, arranged according to their fusibility, may be found at the end of this chapter.

The two large groups are divided into classes according to their fusibility; these again into divisions, etc., by which means we obtain the following general classification:

GROUP I.—MINERALS WITH METALLIC LUSTRE.

CLASS I.—Native malleable metals, and mercury.

CLASS II.—Fusibility 1—5, or readily volatile.

Division 1. Give a strong arsenical odor on Ch.

Division 2. Give on Ch., or in an open tube, the horse-radish odor of selenium.

Division 3. Give in an open tube a white or grayish sublimate, which is fusible into colorless drops, indicative of tellurium. On charcoal a white coating, and coloring the R. Fl. flame green; in presence of selenium, greenish-blue.

CLASS II.—Fusibility 1—5, or readily volatile.

Division 4. Give antimonial fumes on Ch.

Division 5. Give with Sd. on Ch. a sulphur reaction, but do not give indications as above.

Division 6. Do not exhibit the properties of the preceding divisions.

CLASS III.—Infusible, or fusibility above 5, and not volatile.

Division 1. Give with Bx., in very small quantities, in the oxidizing flame, the manganese reaction.

Division 2. Treated on Ch. in R. Fl., become magnetic.

Division 3. Resembling those of division 2.

GROUP II.—MINERALS WITHOUT METALLIC LUSTRE.

CLASS I.—Easily volatile, or combustible.

CLASS II.—Fusibility 1—5; not, or only partially, volatile.

PART I. Fused with Sd. on Ch. give a metallic globule or magnetic metallic mass.

Division 1. Give with Sd. a globule of silver.

Division 2. Give with Sd. a globule of lead.

Division 3. When moistened with hydrochloric acid, color the flame blue, and give with nitric acid a solution which, on addition of an excess of ammonia, assumes an azure-blue color.

Section 1. Give on Ch. a strong arsenical odor.

Section 2. Give no arsenical odor.

Division 4. Impart to the Bx. bead a sapphire-blue color.

Division 5. When fused in forceps, or on Ch. in R. Fl., give a black or gray metallic magnetic mass.

Section 1. Give on fusion on Ch. a strong arsenical odor.

Section 2. Soluble in hydrochloric acid without leaving a perceptible residue, and without gelatinizing.

Section 3. With hydrochloric acid, gelatinize or decompose, with separation of silica.

Section 4. But little affected by hydrochloric acid.

Division 6. Not belonging to either of the preceding divisions.

PART II. Fused with Sd. on Ch., give no metallic globule, or magnetic metallic mass.

Division 1. After fusion and continued heating on Ch., in the forceps, or on platinum foil, have an alkaline reaction, and change to blue the color of moistened red litmus paper, or brown the color of turmeric paper.

Section 1. Easily and completely soluble in water.

Section 2. Insoluble in water, or soluble with difficulty.

Division 2. Soluble in hydrochloric acid without leaving a perceptible residue; some also soluble in water; not gelatinized by evaporation.

Division 3. Soluble in hydrochloric acid, gelatinizing, especially after partial evaporation.

Section 1. Giving water in a matrass.

Section 2. Giving no water in a matrass, or only a trace.

Division 4. Soluble in hydrochloric acid, with separation of silica, without gelatinizing.

Section 1. Giving water in a matrass.

Section 2. Giving no water in a matrass, or only a trace.

Division 5. Little affected by hydrochloric acid; imparting to the Bx. bead the color of manganese.

Division 6. Not belonging to either of the preceding divisions.

CLASS III. — Infusible, or fusibility above 5.

Division 1. After ignition, moistened with SoCo. and again ignited, assume a bright-blue color.

Section 1. Giving much water in a matrass.

Section 2. Giving little or no water in a matrass.

Division 2. Moistened with SoCo. and ignited, assume a green color.

Division 3. After ignition have an alkaline reaction, and change to brown the color of moistened turmeric paper.

Division 4. Completely soluble, or nearly so, in hydrochloric or nitric acid, without gelatinizing or leaving a perceptible residue of silica.

Division 5. With hydrochloric acid, gelatinize or decompose, with separation of silica.

Section 1. Giving water in a matrass.

Section 2. Giving no water in a matrass, or only a trace.

Division 6. Not belonging to either of the preceding divisions.

Section 1. Hardness below 7.

Section 2. Hardness 7, or above.

GROUP I. — MINERALS WITH METALLIC LUSTRE.

CLASS I. — Native Malleable Metals, and Mercury.

Native Silver, see § 197.

Native Gold and Electrum (alloy of silver and gold), see § 150.

Native Copper, see § 134.

Native Lead, characterized by coating on charcoal (see § 23) and softness; H = 1.5.

Native Platinum, see § 152.

Native Palladium, distinguished from the preceding by being soluble in nitric acid.

Native Iron, see § 154.

Native Mercury, see § 187.

Argentite and Hessite are malleable, for which see Divisions 5 and 6.

CLASS II.—Fusibility 1—5, or readily volatile.

Division 1. Give a strong arsenical odor on charcoal.

Native Arsenic, see § 118.

Dufrenoysite, see § 170; *Tennantite*, see § 139; *Polybasite*, see § 207; *Domeykite*, see § 140.

Binnite, $\frac{3}{2}Cu^2S + As^2S^3$. In the closed tube, give a sublimate of sulphide of arsenic; in the open tube, a crystalline sublimate of arsenous oxide, with sulphurous fumes. Before the Blp. on coal give a faint white coating and odor of arsenic. Fuses to a globule, giving metallic copper with soda. Lustre, metallic; color, black on fresh fracture; streak, cherry-red; brittle. $H = 4.5$. $G = 4.4$.

Enargite, $3Cu^2S + As^2S^5$. In closed tube, decrepitates and gives a sublimate of sulphur. In open tube, gives off sulphurous acid and arsenous oxide, the latter condensing to a sublimate containing often antimonous oxide. The roasted mineral gives a globule of copper with fluxes. Lustre, metallic; color and streak, grayish-black. Brittle. Fracture uneven. $H = 3$. $G = 4.4$.

Algodonite, $(Cu^2)^6As^2$. $H = 4$. $G = 7.6$.

Whitneyite, $(Cu^2)^9As^2$. Less fusible than algodonite, otherwise as in domeykite. Massive. Crystalline; very fine granular. Lustre, dull, but strong metallic where scratched; soon tarnishing.

CLASS II.—Fusibility 1—5; or readily, volatile.
Division 1 — (*continued*).

Color, reddish to grayish-white, becoming brown and black on exposure. Malleable. H = 3.5. G = 8.2.

Smaltite, see § 128; Cobaltite, see § 129.

Skutterudite, $CoAs^2$; *Glaucodot*, $(CoFe) S^2 + (CoFe) As^2$; *Alloclasite*, $2CoS^2 + CoAs^2 + 4BiAs$. Before the Blp., all give a sapphire-blue color to the borax bead. Decomposed with nitric acid, with separation of arsenous oxide forming a red solution. In a concentrated solution of alloclasite water, gives a cloudiness, but not in the others. Smaltite, skutterudite, and glaucodote, heated in a matrass, give a sublimate of metallic arsenic. Cobaltite gives none. The strong acid and dilute solutions of cobaltite and glaucodote give a precipitate with chloride of barium; the solutions of smaltite and skutterudite none or a very small one. Smaltite has no cleavage.

Some varieties contain nickel and resemble chloanthite, in which case the nitric acid solution is green. The nickel varieties are distinguished by decomposing the powdered mineral with nitric acid, neutralizing with ammonia without filtering, and afterward, without diluting, filtering. The filtrate will have a fine blue color.

Compare the following minerals, also native bismuth, which often contain cobalt.

Niccolite, see § 191; *Gersdorffite*, see § 192; *Chloanthite* (var. of *Smaltite*), $NiAs^2$, distinguished from gersdorffite by not giving the reactions for sulphur.

Rammelsbergite, $NiAs^2$ (similar to chloanthite). In closed tube, give a sublimate of metallic arsenic. See also *Corynite*, $NiS^2 + Ni,(Sb, As)S^2$, which before the Blp., on coal, give the smell of arsenic and fumes of antimony. Color, silver-white to steel-gray. See *Ullmannite*.

Arsenical Pyrites, see § 161.

Compare also native bismuth, native antimony, which often contain arsenic, but are easily recognized by the white or yellow coating on coal. *Proustite* often has metallic lustre, but is recognized by its red streak.

Division 2. Give on charcoal, or in an open tube, the horse-radish odor of selenium.

Tiemannite, HgSe, *Lehrbachite*, Hg (PbSe); selenide of mercury and lead yield metallic mercury on being heated with Sd. in a closed glass tube (§ 91); the latter yields a globule of metallic lead on being heated on charcoal with Sd.

Clausthalite, PbSe. Color, lead-gray; volatile without previous fusion, depositing first a slight gray, then a white, and finally a greenish-yellow coating; with Sd. yields with difficulty globules of lead.

Naumannite, AgSe. Color, iron-black; melts readily, and yields with Bx. a globule of pure silver.

Berzelianite, Cu^2Se and *Eucairite*, $Cu^2Se + AgSe$. Color of the former, silver-white; of the latter, lead-gray. Distinguished from the other minerals of this division by giving copper reactions.

Crookesite, $(Cu^2TlAg)Se$, is similar, containing 18

CLASS II. — Fusibility 1—5, or readily volatile.
Division 2—(*continued*).

per cent. of thallium, and coloring the flame bright-green. *Zorgite*, PbSe + CuSe., colors the flame blue.

Division 3. Give in an open tube a white or grayish sublimate, which is fusible into colorless drops, indicative of tellurium, see § 11.

The assay-piece used for this experiment ought not to be very small. It must also be borne in mind that the minerals of this division frequently evolve an odor of selenium, owing to a small percentage of selenium which they contain as adventitious constituent.

The minerals of this division may be subdivided according to their color.

a. Ores of tellurium of tin-white or silver-white color.

Native Tellurium, fuses readily and is volatile without leaving a residue.

Hessite, AgTe, and *Altaite*, PbTe; both soluble in nitric acid; the former yields with Sd. on Ch. a globule of metallic silver.

Some varieties of *Sylvanite*, see § 151.

b. Ores of tellurium of lead-gray or steel-gray color.

Tetradymite, see § 123.

Sylvanite, see § 151.

Nagyagite, Pb., Au., Te., S. Color, blackish lead-gray. Distinguished from the preceding by its solution in nitric acid giving a copious precipitate with sulphuric acid.

Division 4. Give copious antimonial fumes on charcoal (see § 16) with a pure white coating, and give no color to the R. Fl.

The fumes possess sometimes the odor of sulphurous acid or arsenic.

Native Antimony, distinguished by its tin-white color; *Stibnite,* see § 115; *Zinkenite,* see § 170; *Jamesonite,* see § 170; *Bournonite,* see § 170.

The powdered stibnite, on being treated with hydrate of potassa, assumes a yellow color, while the latter three minerals, which are steel-gray, do not change color. Bournonite, on being treated with nitric acid, imparts to the solution a sky-blue color, and gives copper reactions on being treated as described in § 71. *Stylotypite* is similar to bournonite, but no precipitate is formed with sulphuric acid from its solution in nitric acid. Zinkenite and jamesonite are converted into white powders by treatment with nitric acid without imparting a color to the acid; they are distinguished by their hardness, that of zinkenite being 3.5, that of jamesonite, 2.5. The former has no cleavage, while in the latter it is very marked in one direction.

Closely resembling the above in their chemical behavior are the following rare minerals: *Boulangerite,* see § 170; *Geocronite,* see § 170; *Plagionite,* see § 170; *Meneghenite,* see § 170.

Discrasite, see § 198; *Stephanite,* see § 206; some varieties of *Tetrahedrite,* see § 138; *Miargyrite,* AgS, Sb^2S^3. Discrasite does not give a sulphur reaction, all the others do. Tetrahedrite gives a copper reaction on being treated as described in § 73. Miargyrite, streak, dark cherry-red; stephanite, streak, black. Miargyrite and stephenite, hardness, 2.5; tetrahedrite, hardness, 3.5.

CLASS II. — Fusibility 1—5, or readily volatile.
Division 4 — (*continued*).

All the minerals of this subdivision give a globule of silver on being treated as described § 104 or § 105.

Brogniardite, Sb^2S^3, AgS, PbS. Isometric. Treated with nitric acid precipitates sulphate of lead.

Freieslebenite, $5(Pb,Ag)S + 2Sb^2S^3$, behaves in the same way, but is monoclinic. Compare also *Pyrargyrite.*

Chalcostibite [Antimonial Copper], Cu^2S, Sb^2S^3; does not give a globule of silver, but yields a globule of metallic copper on being treated with Sd. on charcoal.

Ullmannite, see § 193; *Berthierite*, see § 116; *Breithauptite*, Ni, Sb. All yield a magnetic globule with continued heat. Breithauptite is distinguished from the other two by not giving a sulphur reaction.

Division 5. Give with Sd. on Ch., a sulphur reaction, but do not give the general reactions of the preceding divisions.

Silver Glance, § 203; *Jalpaite*, § 203; *Acanthite*, § 203.

Galenite, see § 169.

Cinnabar, see § 190.

Alabandite, MnS. Isometric. H = 4—4.5. Color, iron-black; streak, green. Lustre, submetallic. The pulverized mineral evolves sulphuretted hydrogen with hydrochloric acid.

Hauerite, MnS^2. H = 4. Color, brownish-black; streak, brownish-red. Lustre, metallic-adamantine. Yields sulphur on being heated in a matrass. These manganese minerals, mixed with phosphorus and saltpetre and boiled, give a fine violet solution.

Chalcocite, see § 137; *Stromeyerite*, see § 208; *Stannite*, see § 210; *Chalcopyrite*, see § 135; *Bornite*, see § 136; *Cubanite*, Cu^2S, Fe^2S^3; *Wittichite*, $3Cu^2S$, Bi^2S^3; *Aikinite* [Acicular Bismuth], $3Cu^2S, Bi^2S^3 + 2(3PbS, Bi^2S^3)$; *Grünauite*, $Bi^2S^3 + 10Ni^2S^3$; *Cuproplumbite*, Cu^2S, 2PbS. *Pentlandite*, NiS, 2FeS. All these minerals are partially soluble in nitric acid, the solution possessing a sky-blue or green color; on addition of water to the concentrated solution a white precipitate is produced, if the mineral under examination was wittichite, grunauite, or aikinite. [To distinguish these three, add to the acid solution sulphuric acid: a precipitate indicates aikinite; wittichite gives the copper reaction on being treated as described in § 73, grunauite not.] Copper pyrites and cuban are distinguished from the others by their brass-yellow color; purple copper is also characterized by its color. To distinguish the remaining four minerals, make a solution in nitric acid; add sulphuric acid: a precipitate indicates cuproplumbite; if no precipitate is produced, add hydrochloric acid: a precipitate indicates stromeyerite; to distinguish between copper-glance and tin pyrites, see § 137 and § 210.

Millerite, see § 194; *Linnæite*, see § 130; *Pyrite*, see § 158; *Marcasite*, see § 159; *Pyrrhotite*, § 160; *Sternbergite*, S, Ag, Fe. The members of this

CLASS II. — Fusibility 1—5, or readily volatile.
Division 5 — (*continued*).

subdivision fuse to globules which are attracted by the magnet. They are readily distinguished by the characteristics given in Chapter III. Sternbergite, by the treatment described § 104, yields a globule of silver. Marcasite and pyrite can only be distinguished by their crystalline form.

Bismuthinite, see § 125; *Chiviatite*, § 125.

Division 6. Do not exhibit the properties of the preceding divisions.

Amalgam, see § 188; *Arguerite*, § 188.

Native Bismuth, see § 122.

Hematite, see § 156.

Cuprite, see § 142. Often with weak metallic lustre.

Magnetite, see § 157.

Wolframite, MnO, FeO, WO^3. H = 5—5.5. G = 7.1—7.5. Orthorhombic. Lustre, submetallic. Streak, dark reddish-brown to black. Opaque. Sometimes magnetic. Color, dark-grayish or brownish-black. Fusibility, 3. The pulverized mineral on being boiled with aqua regia assumes gradually a yellowish color.

Samarskite, NbO^2, FeO, U^2O^3, YO, WO^3, ThO^2, ZrO^2, MnO, CoO, MgO, CaO, HO. Color, velvet-black. Lustre of surface of fracture, shining, and submetallic. Streak, dark reddish-brown. Fusibility, 4.5. By fusing the pulverized mineral with hydrate of potassa, in a silver crucible, boiling the fused mass in water, it gives a green solution, which is filtered; hydrochloric acid gives a white precipitate. If this is boiled with concentrated hydrochloric acid and tin for a few minutes, an equal volume of water added, it gives a bright blue solution.

Rhodonite, dark varieties, $3MnO.SiO^2$, $3HO$. Yields water on being heated in a matrass. Soluble in hydrochloric acid, with separation of silica. In O. Fl., colors the borax glass amethyst.

Compare *Klipsteinite.*

Some varieties of *Psilomelane,* see § 183.

Fayalite, Lievrite, and *Allanite,* some varieties, see p. 169.

Plattnerite PbO^2. Color, iron-black; lustre, metallic-adamantine; streak, brown; opaque; easily reduced to metallic lead.

CLASS III.—Infusible, or fusibility above 5.

Division 1. Give with borax, in very small quantities in the O. Fl., the manganese reactions.

The members of this division are distinguished from each other principally by their physical properties.

Braunite, see § 182; *Hausmannite,* see § 181; *Psilomelane,* see § 183; *Pyrolusite,* see § 180; *Franklinite,* some varieties, see § 186; *Manganite,* Mn^2O^3, HO. Color, steel-gray to iron-black; streak, dark reddish-brown; hardness, 3—4; yields water in a matrass.

Crednerite, $3CuO$, $2Mn^2O^3$. $H = 4.5$. Lustre, metallic; color, iron-black to steel-gray; streak, brownish-black. Moistened with H. Cl., gives a fine blue before the Blp. Dissolved in H. Cl. with ammonia, a precipitate and blue solution, which is not the case with the foregoing.

CLASS III.—Infusible, or fusibility above 5.
Division 1 — (*continued*).

Compare also *Alabandite* and *Hauerite*.

Division 2. Heated on charcoal in reduction flame, become magnetic.

Hematite, see § 156.

Franklinite, see § 186; *Magnetite*, see § 157.

Titaniferous Iron, see § 162; some varieties of *Rutile* and *Brookite* (see below); some varieties of *Limonite* (§ 155), and *Blende* (§ 212).

Division 3. Minerals resembling those of Division 2.

Chromite, see § 127.

Molybdenite, MoS^2; *Graphite*, C., see § 216. Both very soft; hardness, 1.5. Molybdenite, when heated in the forceps, colors the flame greenish; and gives a sulphur reaction when treated as described in § 107.

Brookite, TiO^2; orthorhombic. *Perofskite*, CaO.TiO^2; isometric. Both give the reaction for titanium as described § 111. Distinguished by crystalline form.

Iridosmine, see § 153.

Tantalite and *Columbite*, MnO, FeO, TaO^5, NbO^5, WO^3, SnO^2; *Yttro-tantalite*, 3(CaO.YO.FeO), (TaO^3, WO^3). The color of these minerals is iron-black; yttro-tantalite loses its color before the Blp. and becomes yellowish or white, that of the others remains unchanged. Acids affect them but little. Tantalite and columbite give the same reaction as samarskite when treated with hydrate of potassa, etc. (See p. 158.)

Compare *Polycrase* and *Aeschynite*.

Uraninite, UO, U^2O^3. Color, usually velvet-black; lustre, greasy; partially soluble in nitric acid to a yellow liquid; the solution gives a sulphur-yellow precipitate with ammonia. Boiled with phosphoric acid gives an emerald-green solution.

GROUP II. — MINERALS NOT POSSESSING METALLIC LUSTRE.

CLASS I. — Easily volatile or combustible.

Native Sulphur, S. Completely volatile; burns with a blue flame and emission of sulphurous acid. Color, sulphur-yellow, honey-yellow, and grayish and brownish from impurities.

Realgar, see § 119; *Orpiment*, see § 120.

Arsenolite, see § 121; *Kermesite*, see § 117.

Valentinite, SbO^3. Orthorhombic. Color, white; streak, white; does not change color with hydrate of potassa; does not evolve sulphuretted hydrogen with hydrochloric acid. Lustre, adamantine.

Senarmontite, SbO^3. Isometric. Lustre, resinous, inclining to sub-adamantine; streak, white.

Sal-ammoniac, NH^4Cl; *Mascagnite*, $NH^4O, SO^3 + HO$. Color, white; both evolve ammonia with hydrate of potassa; the former is volatile without previous fusion, the latter intumesces.

Cinnabar, see § 190; *Calomel*, see § 189.

Cotunnite, PbCl. Color and streak, white. Fuses easily on coal; volatilizes, and gives a white coat, the inner edge of which is tinged yellow; with soda, globules of metallic lead.

CLASS II.—Fusibility 1—5; not, or only partially, volatile.

PART I. Give with carbonate of soda on charcoal a metallic globule or a magnetic metallic mass.

Division 1. Give with carbonate of soda a globule of silver.

Proustite, see § 205; *Pyrargyrite*, see § 204; *Xanthoconite*, $3AgS + As^2S^5 + 2(3AgS, As^2S^3)$, behaves like proustite, from which it is distinguished by its yellow streak. Compare Myargyrite.

Cerargyrite, see § 199; *Iodyrite*, see § 202; *Embolite*, see § 200.

Selbite, AgO, CO^2, dissolves in nitric acid with effervescence. Color, ash-gray to black.

Division 2. Give with carbonate of soda a globule of lead.

The minerals of this division are all soluble in nitric acid; the solution gives a copious precipitate with sulphuric acid. If dissolved by boiling with caustic potassa, chromate of potassa directly, or on addition of acetic acid, gives an orange precipitate.

Bindheimite, 3PbO, $SbO^5 + 4HO$. Lustre, resinous or dull. Color and streak, white, grayish-yellow. Before the Blp., on coal, gives a coating of lead and antimony, in the matrass water.

Mimetite, $PbCl + 3(3PbO, AsO^5)$; *Hedyphane*, $PbCl + 3(3[PbO.CaO], [AsO^5.PO^5])$. The former completely, the latter partially reduced to metallic lead with evolution of arsenical fumes.

Pyromorphite, see § 175.

Minium, see § 168; *Crocoite*, see § 177; *Phœnicochroite*, 3PbO, $2CrO^3$. *Dechenite*, (PbO, ZnO)VO^3, often with AsO^5. Crocoite and phœnicochroite give the chromium reaction (§ 67). The latter three, on being boiled with hydrochloric acid, give an emerald-green solution; on adding alcohol to the liquid, concentrating by heat, pouring off from the residue, and then adding water: the liquid assumes a sky-blue color if the mineral was dechenite. The streak of crocoite and dechenite is reddish-yellow, and that of phœnicochroite brick-red.

Linarite, $PbO.SO^3 + CuO.HO$ is characterized by its deep azure-blue color. The color is destroyed by digesting with nitric acid, and sulphate of lead is precipitated.

Cerussite, see § 172; *Phosgenite,* see § 171; *Leadhillite,* see § 173. *Susannite* is of similar composition, but is hexagonal. *Lanarkite,* $PbO.CO^2 + PbO.SO^3$. All soluble in nitric acid with effervescence; leadhillite and lanarkite leave an insoluble residue of lead. The solution of phosgenite gives with nitrate of silver a precipitate of chloride of silver.

Mendipite, $PbCl + 2PbO$; *Matlockite,* $PbCl + PbO$. Dissolve in nitric acid without effervescence; the solution gives a precipitate with nitrate of silver. Colorless; white

Anglesite, see § 174.

Wulfenite, see § 179.

Stolzite, PbO, WO^3. Color, yellow, yellowish-brown; lustre, resinous. Soluble in abundant quantity of hydrochloric acid, leaving a yellowish-green residue (WO^3). With sulphuric acid the pulverized mineral assumes a bright lemon-yellow color.

Vauquelinite, see § 178; *Vanadinite,* $3 PbO, VO^3$ with PbCl. Hexagonal. Color of the former blackish-green, olive-green; of the latter brown,

CLASS II.—Fusibility 1—5; not, or only partially, volatile.
Division 2—(*continued*).

yellowish. Both impart to the borax bead an emerald-green color. Both are soluble in nitric acid. The solution of vanadinite is yellow, and gives a precipitate with nitrate of silver. That of vauquelinite not. *Descloizite*, 2 PbO, VO^3. Orthorhombic.

Division 3. When moistened with hydrochloric acid, color the flame blue; and give with nitric acid a solution which, on addition of an excess of ammonia, assumes an azure-blue color.

Section 1. Give on charcoal a strong arsenical odor.

Chenevixite, Fe^2O^3, $AsO^5 + 3CuO$, HO. Lustre, vitreous. Color, dark-green. Streak, yellowish-green. Fuses to a black magnetic slag, while the following do not.

Bayldonite, 4(PbO, CuO)AsO^5 + 2 HO with PbO. Lustre, resinous. Color, green. Dissolved in nitric acid, gives a precipitate with sulphuric acid of sulphate of lead.

Olivenite, see § 147.

Tyrolite, see § 148; *Chalcophyllite*, $8CuO.AsO^5$ + 12HO. Color, green. Both decrepitate violently and yield much water; chalcophyllite dissolves in ammonia without leaving a residue.

Conichalcite, 3(CuO, CaO), (AsO^5, PO^5) + CuO, HO + $\frac{1}{2}$HO. Fused, gives an alkaline reaction.

Liroconite, AsO^5, PO^5, CuO, Al^2O^3, HO. Color, sky-blue. Does not decrepitate; loses 22 per cent. of water on ignition.

Euchroite, $4CuO.AsO^5 + 7HO$; *Erinite*, 5CuO.

$AsO^5 + 2HO$. Color of both emerald-green. The former loses by ignition 18½ per cent. of water, the latter only 5 per cent. Erinite, amorphous. *Cornwallite*, also amorphous, contains 13 per cent. of water.

Section 2. Do not give an arsenical odor on charcoal.

Atacamite, see § 141. *Tallingite* and *Percylite*.

Chalcanthite, see § 145; *Brochantite*, $2(3CuO.SO^3) + CuO, HO + 4HO$; *Covellite*, CuS. These three minerals give a sulphur reaction (§ 107); chalcanthite is soluble in water, the other two not. Color of covellite, dark indigo-blue; of brochantite, emerald-green, with 12 per cent. of water. *Langite*, greenish-blue color, with 16 per cent. of water.

Cuprite, see § 142; *Melaconite*, CuO. Color, dark steel-gray to black. Both dissolve readily in acids without effervescence (except impure varieties of melaconite).

Malachite, see § 143; *Azurite*, see § 144; *Mysorin*, $CuO.CO^2$. Color, blackish-brown; does not yield water in a matrass. All three dissolve readily in acids with effervescence. *Aurichalcite* gives a zinc coating on coal. *Atlasite*, dissolved in nitric acid, gives a precipitate of chloride of silver with nitrate of silver.

Pseudomalachite, see § 146; *Libethenite*, $4CuO.PO^5 + HO$; *Lunnite*, $6CuO.PO^5 + 3HO$; *Ehlite*, $5CuO.PO^5 + 3HO$; *Tagilite*, $4CuO.PO^5 + 3HO$. Are all readily soluble in nitric acid without effervescence; the (slightly acid) solution gives a precipitate with acetate of lead. Pseudomalachite loses 14 per cent. of water on ignition, the

CLASS II.—Fusibility 1—5; not, or only partially, volatile.
Division 3—(*continued*).

others less (from 7 to 10½). Libethenite is dark olive-green; ehlite, and tagilite, emerald-green.

Torbernite, $2(U^2O^3)$ $PO^5 + CuO.HO + 7HO$. Color, emerald-green. Dissolves in nitric acid to a yellowish-green liquid; on addition of ammonia in excess, a bluish-green precipitate is formed, the supernatant liquid being blue.

Volborthite, $(CuO, CaO)^4, VO^3 + HO$.

Division 4. Impart to the borax bead a blue color.

Erythrite, see § 131; *Annabergite*, see § 196.

Division 5. When fused in the forceps or on charcoal, in reduction flame, give a black metallic magnetic mass.

To observe well the magnetic character of the fused mineral, it is advisable to expose a pretty large assay-piece to the action of the reduction flame.

Section 1. Evolve a strong arsenical odor on being fused on coal.

Scorodite, see § 166; *Pitticite*, Fe^2O^3, $AsO^5 +$ Fe^2O^3, $SO^3 + 15HO$; *Beudantite*, Fe^2O^3, PbO, CuO, HO, SO^3, PO^5, AsO^5. The pulverized minerals assume with hydrate of potassa a reddish-brown color. Scorodite and beudantite occur crystallized; the first orthorhombic, and the second rhombohedral. Their color is usually some shade of green — to brown and black. Streak, greenish-gray to yellow. Lustre, vitreous. Pitticite, massive and reniform. $H = 2—3$. Lustre, vitreous, sometimes greasy. Color, yellowish, brownish, blood-red, and white. Streak, yellow-white. Translucent-opaque.

Arseniosiderite, $6CaO, AsO^5 + (4Fe^2O^3, AsO^5) + 15HO$. Color, yellowish-brown; fibrous; lustre, silky.

Morenosite, $NiO, SO^3 + 7HO$. Partly soluble in water; the solution assumes a blue color on addition of ammonia.

Section 2. Soluble in hydrochloric acid without leaving a perceptible residue, and without gelatinizing. Give no arsenical odor when fused on coal.

Pettkoite, $3FeO, Fe^2O^3 + 2SO^3$, with 1.5 per cent. water. $H = 2.5$. Isometric. Lustre, bright. Color, pure black. Streak, dirty-greenish. Taste, sweetish.

Melanterite, see § 164; *Botryogen*, $3FeO, 2SO^3 + 3(Fe^2O^3, 2SO^3) + 36HO$. Melanterite and botryogen are soluble in water, the latter leaving a yellow residue. The solutions give precipitates with chloride of barium; also with ammonia. Streak of melanterite is green; of botryogen, yellow. *Rœmerite*, yellowish-brown. *Coquimbite*, *Jarosite*, and *Fibroferrite*, all yellow. The last, fibrous and silky, behaves similarly to botryogen. *Voltaite* is distinguished from the foregoing by its black or dark-green color, resinous lustre, and octahedral crystallization. All these sulphates, when heated in the closed tube, give much water.

Siderite, see § 163.

Hureaulite, $5(MnO, FeO) PO^5 + 5HO$. $H = 5$.

Triplite, $3(FeO, MnO)PO^5 + (CaMgFe)F$. H

CLASS II.—Fusibility 1—5; not, or only partially, volatile.
Division 5 — (*continued*).

= 4—5.5. Fuse readily; moistened with sulphuric acid give the phosphoric acid reaction (§ 35); with borax strong manganese reaction; hureaulite yields much water; triplite none, or very little.

Triphylite, 3[FeO, LiO, MnO], PO^5, shows a similar behavior; the manganese reaction is less decided. On dissolving the mineral in hydrochloric acid, evaporating the solution to dryness, adding alcohol, heating the alcohol to ebullition and burning the vapor, the flame assumes a purple color. *Zwieselite*, a clove-brown variety.

Diadochite, $3Fe^2O^3$, $2PO^5+2Fe^2O^3$, $2SO^3+32HO$. H = 3. G = 2.03. Reniform or stalactitic. Lustre, resinous to vitreous. Color, yellowish-brown. Streak uncolored. Soluble in hydrochloric acid. When ignited, gives off sulphuric acid.

Vivianite, see § 165; *Dufrenite*, $2Fe^2O^3.PO^5+3HO$; *Cacoxenite*, $2Fe^2O^3.PO^5+12HO$; *Borickite*, $5(3CaO, Fe^2O^3)2PO^5+15HO$. Fuse readily, and behave with sulphuric acid like the preceding; give no manganese reaction. Yield much water in a matrass: cacoxenite, 33 per cent.; vivianite, 28 per cent.; borickite, 19 per cent.; dufrenite, 8½ per cent. Color of dufrenite, leek-green; of cacoxenite, ochre-yellow; of vivianite, various shades of blue; of borickite, reddish-brown. *Beraunite* is a similar phosphate of red color.

Hematite, see § 156.

Section 3. With hydrochloric acid gelatinize, or are readily decomposed with separation of silica.

Cronstedite, $(\frac{1}{2}(FeO.MnO)^3 + \frac{1}{2}Fe^2O^3)SiO^2 + \frac{3}{2}HO$. $H = 3.5$. Rhombohedral, also amorphous. Color, black; streak, dark leek-green; yields water; gelatinizes with hydrochloric acid. *Sideroschisolite* is probably a variety.

Stilpnomelane, $(3FeO(Al^2O^3, Fe^2O^3)3SiO^2 + 2HO$. $H = 3.4$. *Chalcodite* is similar in composition, often in velvety coatings of brass-like lustre. The color of these minerals is black, yellowish, and greenish-brown. Their streak is greenish-gray.

Voigtite and *Ekmannite* are closely related to the above. They are mica-like in aspect and structure.

Palagonite, of brownish-yellow color and streak; amorphous; yields water and fuses to a black magnetic glass. See also *Jollyte*.

Ilvaite, $\frac{3}{5}(CaO, FeO)^3 + \frac{2}{5}(Fe^2O^3, Al^2O^3)^2\ 2SiO^2$; *Allanite*, $\frac{1}{2}(CeO, FeO, LaO, DiO, YO)^3 + \frac{1}{2}(Al^2O^3, Fe^2O^3)^2 3SiO^2$. Yield no water, or only a trace; gelatinize with hydrochloric acid; allanite fuses with intumescence to a voluminous brownish or blackish glass; ilvaite intumesces but slightly, decrepitates, and fuses to an iron-black bead. Hardness of each, 5.5—6.

Fayalite, $2FeO, SiO^2$. $H = 6.5$. Lustre, metalloid; somewhat resinous in fracture. Color, black, greenish, or brownish-black. Easily fusible, gelatinizes, and attractable by the magnet.

Pyrosmalite, $(\frac{1}{3}HO + \frac{2}{3}(FeO, MnO, FeCl))^2SiO^2$, and

CLASS II.—Fusibility 1—5; not, or only partially, volatile.
Division 5—(*continued*).

Astrophyllite, containing titanium and sometimes zirconium, are decomposed by hydrochloric acid, with separation of silica, without gelatinizing. Fusibility, 2—2.5 *Pyrosmalite* gives the chlorine reaction (§ 65); astrophyllite, not. The hydrochloric acid solution of the latter gives the reaction for titanic acid.

Lepidomelane, $\frac{1}{4}(KO, FeO)^3 + \frac{3}{4}(Al^2O^3, Fe^2O^3)^2 3SiO^2$. H = 3. Lustre, adamantine. Color, black, with occasionally a leek-green reflection. Streak, grayish-green. Easily decomposed by hydrochloric acid, depositing silica in scales.

Andradite [Lime Iron Garnet], $\frac{1}{2}(CaO)+\frac{1}{2}(Fe^2O^3, Al^2O^3)^2 3SiO^2$. Gelatinizes imperfectly; fuses readily; distinguished from the preceding by absence of cleavage. Color, green, brown, black.

Gillingite; Xylotile [a variety of *Serpentine*]. Fuse with difficulty; do not gelatinize. The former is black, amorphous; the latter brown, fibrous, woody. Both yield water in a matrass.

Some impure varieties of *Limonite*, see § 155.

Section 4. But little affected by acids.

Crocidolite, $6(NaO, MgO, FeO), 5SiO^2 + 2HO$; *Arfvedsonite*, $(\frac{2}{5}(\frac{1}{2}FeO+\frac{1}{2}NaO)^3+\frac{3}{5}Fe^2O^3)3SiO^2$. Fusibility, 1.7—2. Color of crocidolite, lavender-blue or leek-green; fibrous; yields water in a matrass; arfvedsonite is black, and yields no water.

[See also *Hornblende* and *Tourmaline*, below, some varieties of which become slightly magnetic after fusion.]

Celadonite [Green Earth]. Fusibility, 3; color, celandine-green; hardness, 1; earthy.

Acmite, $(\frac{1}{3}(FeO, NaO)^3 + \frac{2}{3}Fe^2O^3)3SiO^2$; *Babingtonite,* $(\frac{3}{4}(CaO, FeO, MnO)^3 + \frac{1}{4}Fe^2O^3)3SiO^2$. Fusibility of the former, 2; of the latter, 2.6; form a black lustrous slag. Both are cleavable.

Almandite [Iron Garnet], $(\frac{1}{2}(FeO)^3 + \frac{1}{2}(Al^2O^3))^2 3SiO^2$. Fusibility, 3; hardness, 7—7.5. Color, reddish-brown. Not cleavable. See also *Allochroite.*

Wolframite and *Ferberite,* WO^3, FeO, MnO. Color, black; streak, brownish; metallic vitreous lustre; boiled with concentrated phosphoric acid, a blue syrup, which, diluted with water, becomes colorless. If powdered iron is added and then shaken, it gives a fine blue color.

Megabasite, mainly MnO, WO^3, behaves in a similar manner. Streak, ochre-yellow.

Rhodonite, MnO, SiO^2. $H = 5.5—6$. Color, brownish-red to flesh-red. Streak, white. Often magnetic after fusion. Colors borax bead amethyst.

Lepidolite, often becomes magnetic on fusing, and colors the flame reddish-purple. Compare *Epidote.* Color, pistachio-green.

Stilpnomelane. With difficulty decomposed by hydrochloric acid. Fusibility, 3. Color, black. Yields water in a matrass. Compare *Lepidomelane.*

Division 6. Not belonging to either of the preceding divisions.

Molybdite, MoO^3. Color, sulphur-yellow; earthy. Gives with the fluxes the reactions of molybdic

CLASS II.—Fusibility 1—5; not, or only partially, volatile.
Division 6—(*continued*).

acid. Dissolves readily in hydrochloric acid; the solution is colorless, but turns blue on being stirred with an iron spatula.

Eulytite, $4BiO^3$, $9SiO^2$. Fuses easily to a brown bead. Gelatinizes with hydrochloric acid. On charcoal, with soda, yields a globule of metallic bismuth. Color, brown to yellow.

Bismutite, see § 124.

PART II. With carbonate of soda on charcoal, give no metallic globule or magnetic metallic mass.

Division 1. After fusion and continued heating on charcoal in the forceps or in the platinum spoon, have an alkaline reaction, coloring moistened turmeric paper reddish-brown, and change to blue the color of a moistened red litmus paper. The test may be made with splinters and not with the powder.

Section 1. Readily and completely soluble in water.

Nitre, $KO.NO^5$; *Soda Nitre*, $NaO.NO^5$. Deflagrate vividly on burning coals. Fused on platinum wire, the former colors the flame bluish, with a red tint; the latter, bright-yellow.

Natron, $NaO.CO^2 + 10HO$; *Trona*, $2NaO.3CO^2 + 4HO$. The watery solution has an alkaline reaction, and effervesces on addition of hydrochloric acid. Crystals of the former decompose quickly in the air, the latter not.

Mirabalite, $NaO.SO^3 + 10HO$; *Thenardite*, $NaO.SO^3$; *Glaserite*, $KO.SO^3$; *Epsomite*, $MgO.SO^3 + 7HO$; *Kalinite*, $KO.SO^3 + Al^2O^33SO^3 + 24HO$. The watery solutions of these minerals give a copious precipitate with chloride of barium; the solution of potash, alum, and epsomite are precipitated by carbonate of potassa (distinguished by reaction with solution of cobalt, § 44). *Kainite*, $2MgO.SO^3 + KCl + 6HO$, behaves in a similar manner; soluble in water, and a precipitate is formed with nitrate of silver. The concentrated solution of glaserite gives a precipitate with bichloride of platinum; mirabalite yields much water, thenardite none.

Halite, NaCl; *Sylvite*, KCl. The watery solution gives a copious precipitate with nitrate of silver. Gives also the reactions for chlorine described §§ 65, 66. The latter gives a heavy yellow precipitate with solution of platinum, but the former does not.

Borax, $NaO.2BO^3 + 10HO$. Gives the reaction for boracic acid, § 60.

Section 2. Insoluble in water, or soluble with difficulty.

Hayesine, CaO, $2BO^3 + 6HO$. Fusibility, 1; coloring the flame yellow. Yields much water. Moistened with sulphuric acid, the flame changes momentarily to green. Somewhat soluble in hot water, giving alkaline reaction. The reactions of *Ulexite* are the same.

Gay-Lussite, $CaO.CO^2 + NaO.CO^2 + 5HO$; *Witherite*, $BaO.CO^2$; *Staffelite*, $3CaO.PO^5 + CaO.CO^2$. Dissolve in dilute hydrochloric acid with effervescence; the first yields water, the latter do not. The solution of the staffelite gives a precipitate with ammonia, the others not. Compare *Strontianite*, which colors the flame crimson.

CLASS II.—Fusibility 1—5; not, or only partially, volatile.
Division 1—(*continued*).

Anhydrite, $CaO.SO^3$; *Gypsum*, $CaO.SO^3 + 2HO$; *Polyhalite*, $KO.SO^3 + MgO.SO^3 + 2(CaO.SO^3) + 2HO$; *Glauberite*, $NaO.SO^3 + CaO.SO^3$. Soluble in much hydrochloric acid; in the solution chloride of barium gives a precipitate. Gypsum yields much water, polyhalite little, the rest none; anhydrite is distinguished by superior hardness, 3.5; polyhalite is distinguished from glauberite by its solution giving a yellow precipitate with bichloride of platinum.

Barite, $BaO.SO^3$; *Celestite*, $SrO.SO^3$. Insoluble in hydrochloric acid; give a sulphur reaction when treated as described § 107. Celestite colors the flame red, § 34; barite yellowish-green, § 35.

Fluorite, CaF; *Cryolite*, $3NaF + Al^2F^3$; *Pharmacolite*, $2CaO.3AsO^5 + 6HO$. Do not effervesce with acids, and give no sulphur reaction. Pharmacolite evolves arsenical odor on charcoal; the other two give fluorine reaction, § 76. Fusibility of fluorite, 3; of cryolite, 1. *Antozonite*, var. of fluorite, gives odor of antozone.

Chiolite, $3NaF + 2Al^2F^3$, behaves like cryolite; occurs only massive-granular; while cryolite is distinctly crystalline, and cleavable in three directions.

Pachnolite yields strongly acid water. Closely related are *Arksutite* and *Chodneffite*, without water; *Thomsenolite* and *Gearksutite* with water.

Cancrinite, $(\frac{1}{4}(NaO, KO, CO)^3 + \frac{3}{4}Al^2O^3)^2\ 3SiO^2 + \frac{3}{4}SiO^2 + NaO.CO^2$. Effervesces with hydrochloric acid, and gelatinizes. In the flame it grows white and opaque, and then melts (2.5), intumesces, and forms a white blebby mass. The easy fusibility distinguishes it from nephelite.

Division 2. Soluble in hydrochloric acid without leaving a perceptible residue; some also soluble in water; not gelatinizing.

Tschermigite, $NH^4O.SO^3 + Al^2O^3.3SO^3 + 24HO$; *Alunogen*, $Al^2O^3.3SO^3 + 18HO$; *Goslarite*, $ZnO.SO^3 + 7HO$. All soluble in water; give sulphur reaction, § 107. Heated on charcoal and treated with solution of cobalt, the former assume a blue, the latter a green color, §§ 44, 45. The first with caustic potassa gives the smell of ammonia, the second does not.

Chondrarsenite, $5MnO, AsO^5 + 2\frac{1}{2}HO$. Easily fusible, giving arsenic fumes on coal, and amethyst color to the borax bead. Color, yellow.

Adamite, $3ZnO, AsO^5 + ZnO.HO$. Easily fusible, giving arsenic fumes on coal, with a coating of zinc. Color, honey-yellow.

Struvite, $NH^4O, 2MgO, PO^5 + 12HO$. Melts easily. Yields water in matrass, with caustic potassa ammonia, and with hydrochloric acid fumes of chloride of ammonium.

Sassolite, $BO^3, 3HO$; *Boracite*, $3MgO, 4BO^3$; *Hydroboracite*, $3CaO, 4BO^3 + 3MgO, 4BO^3 + 18HO$. Give the boracic acid reaction, § 60. Sassolite is soluble in alcohol, the others not; boracite yields no water, while the others do. Hydroboracite contains 26 per cent. water, and a similar mineral, szaibelyite, 7 per cent. *Stassfurtite* is closely related. Compare borax.

CLASS II.—Fusibility 1—5; not, or only partially, volatile.
Division 2 — (*continued*).

Alabandite, MnS, and *Hauerite*, MnS^2, give strong manganese reaction; see p. 115.

Wagnerite, MgF + $3MgO.PO^5$; *Apatite*, 3CaO.PO^5 + $\frac{1}{2}$Ca(Cl,F). Moistened with sulphuric acid, impart a pale bluish-green color to the flame. Fusibility of wagnerite, 3—3.5 (with intumescence); of apatite, 5 (without intumescence); wagnerite is soluble in dilute sulphuric acid; apatite not.

Brushite $(\frac{2}{3}CaO + \frac{1}{3}HO)^3PO^5 + 4HO$. Behaves in the wet way like apatite, but yields 26 per cent. water.

Amblygonite $[\frac{1}{4}(LiO,NaO)^3 + \frac{3}{4}Al^2O^3]^4PO5$. Fusibility, 2; hardness, 6. With difficulty soluble in concentrated sulphuric or hydrochloric acid.

Torbernite $(2U^2O^3)PO^5 + CuO.HO + 7HO$. *Autunite* $(2U^2O^3)PO^5 + CaO.HO + 7HO$. Fuse readily, yield water, and give with fluxes the reactions of sesquioxide of uranium. See Table II. Soluble in nitric acid. The first gives a globule of copper with soda on coal.

Division 3. Soluble in hydrochloric acid, forming a perfect jelly.

Section 1. Give water in a matrass.

Datolite $(3CaO,3HO,BO^3)SiO^2$. Yields but little water, and gives the boracic acid reaction, § 60.

Edingtonite, SiO^2,Al^2O^3,BaO,etc. The dilute hydrochloric acid solution gives a precipitate, with sulphuric acid, of sulphate of baryta. Sp. gr. 2.7.

Natrolite, $NaO,Al^2O^3,3SiO^2+2HO$. Fusibility, 2; does not intumesce; hardness, 5—5.5.

Scolecite, $CaO,Al^2O^3,3SiO^2 + 3HO$; *Laumontite* $(\frac{1}{4}(3CaO) + \frac{3}{4}Al^2O^3)3SiO^2 + 3HO$. Scolecite, on being heated, curls up like a worm, and finally melts to a bulky, shining slag, which in the inner flame becomes a vesicular, slightly translucent bead; hardness, 5.5. Pyroelectric. Laumontite intumesces and fuses to a white translucent enamel; hardness, 3.

Nearly related to Scolecite, and showing a similar behavior, are *Mesolite* and *Thomsonite*, but they are not pyroelectric.

Phillipsite $(\frac{2}{3}CaO + \frac{1}{3}KO)Al^2O^3,4SiO^2,5HO$.H, 4.5. Fusibility 3, with slight intumescence; occurs always in twin crystals. Lustre vitreous. Color white, sometimes reddish. *Gismondite* is closely related. Orthorhombic, with forms often resembling square octahedrons. H = 4.5. Lustre splendent. Compare, in div. 4, *Apophyllite*, *Okenite*, and *Analcite*, which gelatinize with hydrochloric acid.

Section 2. Giving only traces or no water in a matrass.

Helvite $(\frac{1}{2}(MnO,FeO) + \frac{1}{2}BeO)^2SiO^2 + \frac{1}{3}MnS$; *Tephroite*, $2MnO,SiO^2$. Distinguished from the other minerals of this section by giving manganese reactions. Color of helvite, wax-yellow; hardness, 6—6.5; of tephroite, ash-gray; hardness, 5.5—6. *Danalite*, containing zinc, gives with soda on coal a small slag of zinc, and with borax the iron reaction. Color, flesh-red to gray.

Hauynite and *Lapis Lazuli*, $SiO^2Al^2O^3,CaO,NaO,SO^3,S$, are of azure-blue color; give sulphur re-

CLASS II.—Fusibility 1—5; not, or only partially, volatile.
Division 3 — (*continued*).

action, § 107. Fusibility of the former, 4.5; of the latter, 3, forming a white glass.

Nosite and *Skolopsite*, $SiO^2,Al^2O^3,CaO,NaO,SO^3$, of gray or brownish color; give sulphur reaction, § 107. Fusibility of nosite, 4.5; of skolopsite, 3 (with intumescence like idocrase). The former crystallizes in dodecahedrons, the latter occurs granular-massive.

Sodalite, $2NaCl + (3NaO)^2(SiO^2)^3 + 3((Al^2O^3)SiO^2)^3)$; *Eudialyte*, $2(CaO,NaO)^3,2SiO^2+ZrO^2,2SiO^2$, give the chlorine reaction, § 65. The former fuses to a transparent, colorless glass, the latter to a grayish-green scoria or opaque glass. The dilute hydrochloric solution of eudialyte colors the turmeric paper orange-yellow; boiled with sulphate of potassa, and evaporated to crystallization and then boiled with water, a precipitate of zirconia is formed which makes the solution cloudy.

Wollastonite, CaO,SiO^2. Fuses quietly to a colorless, semi-transparent glass. The hydrochloric acid solution gives no, or only a very slight, precipitate with ammonia; but with the carbonate a bulky precipitate. See also *Pectolite*.

Nephelite $(\frac{1}{4}(NaO,KO,CaO)^3 + \frac{3}{4}Al^2O^3)^2 3SiO^2 + \frac{3}{4} SiO^2$.

Melilite $(\frac{2}{3}(NaO, MgO, CaO)^3 + \frac{1}{3}(Al^2O^3, Fe^2O^3)^2) 3SiO^2$.

Meionite $(\frac{1}{3}(\frac{10}{11}CaO+\frac{1}{11}NaO)^3+\frac{2}{3}Al^2O^3)^2 3SiO^2$.

With the solution of these minerals in hydrochloric acid, ammonia gives a precipitate. Meionite fuses with intumescence, the others quietly. Nephelite hexagonal, melilite tetragonal (compare *Cancrinite*). The behavior of *Barsowite* is similar to melilite, but fuses with more difficulty, and quietly. (Compare *Gehlenite,* which is nearly infusible, and *Zachylite.*)

Division 4. Soluble in hydrochloric acid, with separation of silica, without forming a perfect jelly. (It is sometimes necessary to treat the finely-pulverized mineral with concentrated acid.)

Section 1. Giving water in a matrass.

Klipsteinite, $(MnO)^3SiO^2 + (2Mn^2O^3)SiO^2 + 4HO$. Easily decomposed by hydrochloric acid, evolving chlorine. Fuses to a black slag in the oxidizing flame. With phosphoric acid it gives a violet solution.

Apophyllite, $(HO,KO,CaO)^2SiO^2 + HO,SiO^2$. *Pectolite* $(\frac{4}{6}CaO + \frac{1}{6}NaO + \frac{1}{6}HO)SiO^2$. *Okenite,* $(\frac{1}{2}CaO + \frac{1}{2}HO)SiO^2 + \frac{1}{2}HO$. The silica separates in the shape of gelatinous lumps. The hydrochloric acid solution gives no, or only a slight, precipitate with ammonia. Pectolite yields but little water, the others much. Fusibility of apophyllite, 1.5, forming a white vesicular glass; of okenite, 2.5—3, forming a porcelain-like mass. (Compare *Xonaltite.*)

Analcite, $NaO,Al^2O^3,4SiO^2,2HO$. Gelatinizes like the preceding; in the acid solution ammonia produces a copious precipitate.

Pyrosclerite $(\frac{2}{3}(MgO)^3 + \frac{1}{3}Al^2O^3)^2SiO^2 + 3HO$. *Chonicrite,* $CaO,MgO,Fe^2O^3,Al^2O^3,SiO^2,HO$. *Jollyte,* $(\frac{1}{3}(FeO,MgO)^3 + \frac{2}{3}Al^2O^3)^2 3SiO^2 + 4HO$. Are distinguished from the other minerals of this section

CLASS II.—Fusibility 1—5; not, or only partially, volatile.
Division 4—(*continued*).

by their inferior hardness, 2.5—3. Chonicrite fuses from 3.5—4, with intumescence; has no cleavage; whitish. Fusibility of pyrosclerite 4, without intumescence; cleavable in one direction; green. Jollyte fuses with difficulty; amorphous; brown; powder light-green.

Brewsterite, $(\frac{2}{3}SrO,\frac{1}{3}BaO),Al^2O^3,6SiO^2,5HO$; characterized by its hydrochloric acid solution giving a precipitate with sulphuric acid.

Stilbite, $CaO,Al^2O^3,6SiO^2,6HO$. *Hypostilbite*, $(\frac{7}{9}CaO + \frac{2}{9}NaO),Al^2O^3 + 4\frac{1}{2}SiO^2),6HO$. *Chabazite*, $\frac{4}{5}CaO+\frac{1}{5}(NaO,KO),Al^2O^3,4SiO^2,6HO$. *Prehnite* $(\frac{2}{6}CaO+\frac{3}{6}Al^2O^3+\frac{1}{6}(3HO)^2,3Sio^2$. Fuse with intumescence to enamel-like masses. Prehnite yields but little water, losing by ignition only 4.3 per cent.; the others lose from 15 to 20 per cent. Chabazite is distinguished by its rhombohedral crystallization and imperfect cleavage. In stilbite and hypostilbite the cleavage is perfect in one direction. Stilbite is orthorhombic, and hypostilbite occurs in radiate-fibrous or columnar masses. *Mordenite*, H=5, occurs in hemispherical, reniform, or cylindrical concretions, with a fibrous structure. Yields 12 per cent. water, and fuses without intumescence. Not perfectly decomposed by acids.

Mosandrite, $(Ce,CaO,NaO,SiO^2,TiO^2,HO)$, and *Catapleiite*, have hardness 4 and 6, and distinct cleavage. The first fuses quietly to a yellowish-brown glass; fusibility 2.5; *Catapleiite*, 3, giving a white porcelain bead. It is soluble in hydrochloric acid without gelatinizing, and gives the zirconia reaction, coloring turmeric paper orange-yellow.

Sepiolite [Meerschaum] $(\frac{2}{3}MgO+\frac{1}{3}HO)SiO^2+\frac{1}{3}HO$, see below; *Deweylite*, $(\frac{2}{3}MgO+\frac{1}{3}HO)SiO^2+\frac{4}{3}HO$. Distinguished by being much less fusible than the preceding (fusibility, 5); the former absorbs water with great avidity, the latter not.

Sordavalite $(\frac{1}{2}(MgO,FeO)^3+\frac{1}{2}Al^2O^3)3SiO^2$. Amorphous. Fusibility, 2.5, forming a thick, black, brilliant glass. Color, brownish-black.

Section 2. Giving only traces or no water in a matrass.

Cryophyllite. Micaceous. Fuses easily in the flame of a candle, giving the flame a lithia reaction.

Tachylyte, KO,NaO,CaO,MgO,FeO, etc. Fuses readily to a black, shining glass. Hardness, 6.5; color, black.

Schorlomite, $(\frac{4}{11}(CaO)^3 + \frac{3}{11}Fe^2O^3 + \frac{4}{11}(\frac{3}{2}TiO^2))^4 3SiO^2$.

Wernerite $(\frac{1}{3}(CaO,NaO)^3+\frac{2}{3}Al^2O^3)^2 3SiO^2$. Color, light; hardness, 5—5.5. Fuses with intumescence to a white, vesicular glass.

Woehlerite $[\frac{2}{3}(NaO,CaO)^2+\frac{1}{3}ZrO]SiO^2[+\frac{1}{10}(FeO,MnO)CbO^5]$. Fusibility 3, forming a yellowish enamel. From the hydrochloric acid solution silica and columbic acid separate. *Eucolite* probably belongs here. With the fluxes gives the reactions of silica, manganese, and iron. Color, wine-, honey-, resin-yellow, brownish-red.

Labradorite $(\frac{1}{4}(NaO,CaO)^3 + \frac{3}{4}Al^2O^3)^2, 3SiO^2+\frac{3}{2}SiO^2$. *Anorthite* $(\frac{1}{4}(CaO)^3+\frac{3}{4}Al^2O^3)^2 3SiO^2$. Fusi-

CLASS II.—Fusibility 1—5; not, or only partially volatile.
Division 4—(*continued*).

bility, 3—4, without intumescence, forming a colorless glass. Hardness of the former, 6; of the latter, 6—7. Cleavage perfect.

Grossularite [some varieties], $(\frac{1}{2}(3CaO) + \frac{1}{2}(Al^2 O^3))^2 3SiO^2$. Fusibility, 3. Not cleavable.

Titanite [Sphene], some varieties, see below. Gives titanium reactions, § 111. See *Danburite*, which gives a fine green flame. Also *Tephroite*, which gives an amethyst color to the borax bead.

Division 5. Little affected by hydrochloric acid; give with fluxes the manganese reactions.

Carpholite $(Al^2O^3, Mn^2O^3, Fe^2O^3)^2 3SiO^2 + 3HO$. Occurs only in radiated and stellated tufts. Color, straw-yellow. Silky. Yields water.

Spessartite [Manganese Garnet], $(\frac{1}{2}(MnO,FeO)^3 + \frac{1}{2}Al^2O^3)^2 3SiO^2$. Color, brownish-red. Fuses without intumescence. Not cleavable.

Piedmontite, $(\frac{1}{3}(CaO)^3 + \frac{2}{3}(Al^2O^3, Mn^2O^3, Fe^2O^3))^2 3SiO^2$. Fusibility, 2—2.5; intumesces. Cleavable. Color, cherry-red to reddish-black.

Rhodonite, MnO, SiO^2. Fusibility, 3, without intumescence. Color, rose-red; cleavable.

Division 6. Not belonging to either of the preceding divisions. All are silicates except *Scheelite*, and are not decomposed, or only partially, by hydrochloric acid.

Danburite, $(\frac{1}{4}(CaO)^3 + \frac{3}{4}BO^3)^2 3SiO^2$. Fusibility, 3, and gives a fine green color to the flame. The bead is clear while hot, and cloudy when cold. *Howlite* is closely related.

Scheelite, CaO, WO^3. Fusibility, 5. Soluble in hydrochloric acid, leaving a residue of tungstic acid, which is soluble in ammonia, and which gives with S. Ph. the characteristic reaction of tungstic acid; see Table II.

Lepidolite, $(3LiO,Al^2O^3)^2\ 3SiO^2 + 2((LiO)^3, Al^2O^3)\ 3SiO^2$, and *Cookeite*, $(SiO, KO, Al^2O^3, SiO^2, HO)$, are micaceous, splitting very easily in one direction. Fusibility of lepidolite is 3, colors the flame crimson, and gives no water in matrass. Cookeite intumesces, colors the flame crimson, but yields much water in matrass.

Thermophyllite, (SiO^2, MgO, HO), *Euphyllite*, and *Margarite*, are all micaceous in structure. The first intumesces before the flame and yields much water. The others fuse without intumescence (4—4.5), and yield little water. Their laminæ are not elastic. Euphyllite is easily decomposed by sulphuric acid, and margarite with difficulty. Compare *Muscovite* and *Biotite*.

Petalite, $(\frac{1}{5}(LiO, NaO)^2 + \frac{4}{5}Al^2O^2)\ 3SiO^2 + 3SiO^2$, and *Spodumene*, $(\frac{1}{5}(LiO, NaO) + \frac{4}{5}Al^2O^3)\ 3SiO^2$, do not possess as perfect a cleavage as the preceding, and greater hardness; hardness of petalite, 6—6.5; of spodumene, 6.5—7. Both give the lithia reaction, § 89. Spodumene fuses with intumescence to a glassy globule; petalite fuses to a white enamel.

Leucophanite fuses easily and quietly to a transparent and colorless glass. Cleavage very marked in one direction. $H = 3.5—4$. If heated, phosphoresces with a reddish-violet light, also if struck with a hammer in the dark.

CLASS II.—Fusibility 1—5; not, or only partially, volatile.
Division 6—(*continued*).

Wilsonite. Fusibility, 2, swelling up to a whitish glass. Yields water in matrass. H = 3.

Sordavallite. Fusibility, 2.5; amorphous; brownish-black. (See Div. 4.)

Diallage, (CaO.MgO), SiO^2. Fusibility, 3.5; characterized by its pearly metallic lustre; cleaves easily in one direction.

Harmotome, BaO,Al^2O^3,5SiO^2,5HO. Distinguished from most of the other minerals of this division by yielding water in a matrass. In the partial solution in hydrochloric acid, sulphuric acid gives a precipitate with the baryta. Occurs usually in twin crystals.

Axinite, (3CaO(Al^2O^3, Fe^2O^3, Mn^2O^3, BeO^3)^2 3SiO^2; *Tourmaline,* (3RO, R^2O^3, BO^3)^8 9SiO^2. (R = NaO, CaO, MgO, FeO;) R^2O^3 = Al^2O^3, Fe^2O^3.) Give the reaction of boracic acid, § 61. Axinite fuses readily with intumescence to a dark-green glass. Different varieties of tourmaline show different fusibility. Hardness of axinite, 6.5; of tourmaline, 7—7.5. Heat develops electricity in tourmaline, but not in axinite.

Pyroxene (var. *Malacolite*), (CaO, MgO)SiO^2, and (var. *Augite*), (CaO, MgO, FeO) (SiO^2, $Al^2O^3\frac{2}{3}$). Their H = 6. Fusibility, 3.5—4. Malacolite fuses to a whitish, augite to a black glass. Color of augite, black or dark-green; of malacolite, pale-green or colorless.

Var. *Hedenbergite,* (CaO, FeO,) SiO^2.

Amphibole (var. *Tremolite*), (CaO, MgO.)SiO^2;

(var. *Actinolite*), (CaO, MgO, FeO,) SiO^2, and (var. *Hornblende*), (MgO, CaO, FeO) (SiO^2, $Al^2O^3\frac{2}{3}$.) Their H = 5.5; fusibility, 3—4. Tremolite fuses to a white or light-colored glass, hornblende to a black or gray glass; the first is colorless or white, or of light-green, yellow, or gray color; hornblende and actinolite are green or black. Some varieties of *Asbestos* and *Amianthus* belong here. *Nephrite* is in part a tough, compact, fine-grained tremolite, having a tinge of green or blue, a splintery fracture, and somewhat greasy feel. H = 6—6.5.

Titanite, (CaO + TiO)SiO^2. Fusibility, 3. H = 5—5.5. Monoclinic. Gives the titanium reaction. § 111. Imperfectly soluble in hydrochloric acid.

Guarinite, of similar composition, but tetragonal.

Keilhauite, containing 28 per cent. of TiO^2. Fuses with intumescence to a black shining glass. Yields with borax an iron-colored glass, which in the inner flame becomes blood-red. Reaction of manganese with soda. Decomposed by hydrochloric acid.

Orthoclase, ($\frac{1}{4}$(3KO) + $\frac{3}{4}Al^2O^3)^2$ $3SiO^2$ + $6SiO^2$. *Albite* with potassa replaced by soda. Hardness, 6. Fuse without intumescence; fusibility of orthoclase, 5; of albite, 4; the latter colors the flame yellow. Not soluble in acids. With solution of cobalt become blue on the edges, § 44.

Oligoclase, ($\frac{1}{4}$(NaO, CaO)^3 + $\frac{3}{4}Al^2O^3)^2 3SiO^2$ + $3\frac{3}{4}$ SiO^2, is more fusible than albite. It sometimes resembles laboradorite, but, unlike it, is not materially acted upon by acids.

Hyalophane is very similar to these minerals, but

CLASS II.—Fusibility 1—5; not, or only partially, volatile.
Division 6—(continued).

if fused with potassa, treated with hydrochloric acid and water, the solution gives a precipitate of baryta with sulphuric acid.

Zoisite, $(\frac{1}{3}(3CaO) + \frac{2}{3}Al^2O^3)^2 3SiO^2$, and *Epidote*, $(\frac{1}{3}(3CaO) + \frac{2}{3}(Al^2O^3, Fe^2O^3))^2 3SiO^2$. Hardness, 6.5. Fusibility, 3—3.5; fuse with intumescence —zoisite to a white or yellowish slag, epidote to a black or dark-brown slag. Color of zoisite, gray, yellowish-gray, grayish-white; of epidote, green.

Garnet (var. *Grossularite*), $(\frac{1}{2}(3CaO) + \frac{1}{2}Al^2O^3)^2$ $3SiO^2$; (var. *Pyrope*), $(\frac{1}{2}(CaO, MgO, FeO, MnO)^3 + \frac{1}{2}Al^2O^3)^2$ $3SiO^2$, and *Vesuvianite*, $(\frac{3}{5}(CaO, MgO, FeO)^3 + \frac{2}{5}Al^2O^3)^2 3SiO^2$. Hardness, 6.5—7.5. Fusibility of lime garnet and vesuvianite, 3; of pyrope, 4.5. Vesuvianite possesses cleavage, the others not. Pyrope gives, with the fluxes, the chromium reactions, is not acted upon by acids, and is of a blood-red color. The others, green, yellowish-brown, hyacinth-red, and white.

Edelforsite (var. of *Wollastonite*). H = 6—7. Fusibility, 4; somewhat intumescent. Not acted upon by acids. Heated, gives a greenish-yellow phosphorescence. *Sphenoclase*, similar in its behavior, but fuses quietly and more easily, and phosphoresces with a faint yellowish light. (See also *Emerald*, *Euclase*, *Iolite*, *Biotite*, and *Muscovite*.)

Obsidian, *Pitchstone*, *Pearlstone*, and *Pumice*, SiO^2, Al^2O^3, NaO, KO, HO, are amorphous. Fusibility, 3.5—4; fuse with intumescence to porcelain-like masses, or white vesicular glasses. Lustre of obsidian glassy, of pitchstone greasy, of pearlstone pearly; pumice is characterized by its porosity.

CLASS III. — Infusible, or fusibility above 5.

Division 1. After ignition, moistened with solution of cobalt and again ignited, assume a bright-blue color. (Some minerals should be first calcined and pulverized.)

With the hard, anhydrous minerals of this division, the color is best seen by reducing the substance to a fine powder and moistening this with the solution of cobalt. The color appears only after cooling, and by daylight. By calcination *Alunite* loses 13 per cent. water, *Aluminite* 47, and a similar mineral, *Felsobanyite*, 37.

Pissophanite, $(Al^2O^3, Fe^2O^3)^5 2SO^3 + 30HO$, blackens in the flame, to which it gives a greenish tinge. Aluminite is white and opaque, while the latter is greenish and transparent.

(See, also, *Alunogen* and *Tschermigite*, which are soluble in water, while the foregoing are not.)

Similar minerals: *Trolleite* contains 6 per cent. water; *Sphærite*, 23; *Tavistockite*, 12.

Gibbsite, $Al^2O^3, 3HO$; *Diaspore*, Al^2O^3, HO; *Seybertite*, (Al^2O^3, CaO, MgO, etc.), and *Pholerite*, $2Al^2O^3, 3SiO^2 + 4HO$. Gibbsite is easily soluble in hydrate of potassa, and loses by ignition 34.5 per cent. water. The others are insoluble in potassa. Distinct cleavage in one direction. Seybertite loses 4½ per cent. water by ignition. Color, wax-yellow. Diaspore and pholerite lose 15 per cent., and may be distinguished from the

CLASS III. — Infusible, or fusibility above 5.
Division 1 — (*continued*).

other minerals by their hardness. Diaspore, 6; pholerite, 1. The last often occurs in scales with a mother-of-pearl lustre.

Section 1. Giving much water in a matrass.

Alunite, $KO.SO^3,(Al^2O^3.SO^3)^3 + 6HO$; *Aluminite*, Al^2O^3, $SO^3 + 9HO$. Give a sulphur reaction, § 107. Aluminite is readily soluble in hydrochloric acid; alunite not visibly affected.

(See, also, *Ammonia Alum*, and *Potash Alum*.)

Plumbo-gummite, see § 176.

Calamine, see § 214.

Wavellite, $3Al^2O^3$, $2PO^5+12HO$; *Evansite*, Al^2O^3, $3HO + 2Al^2O^3$, PO^5+15HO; *Peganite*, $2Al^2O^3$, $PO^5 + 6HO$; *Fischerite*, $2Al^2O^3$, $PO^5 + 8HO$; *Berlinite*, Al^2O^3, $PO^5 + \frac{1}{2}HO$; *Richmondite*, Al^2O^3, $PO^5 + HO$. Soluble to a great extent in hydrate of potassa. Give the reactions of phosphoric acid, §§ 94 and 95. The former two occur usually in globular concretions of radiated structure, the latter two minutely crystalline. Peganite loses on ignition 24 per cent. of water; wavellite, 27; fischerite, 29; richmondite, 35; evansite, 40.

Allophane, Al^2O^3, SiO^2+6HO; *Halloysite*, $(\frac{1}{4}(HO)^3 + \frac{3}{4}Al^2O^3)^2$ $SiO^2 + 3HO$; *Samoite*, $2Al^2O^3$, $3 SiO^2 + 10HO$; *Collyrite*, $2Al^2O^3$, $SiO^2 + 9HO$. Decomposed by hydrochloric acid with separation of gelatinous silica. Hardness of allophane, 3; gelatinizes completely; often colors the flame green, showing the presence of copper, and loses by ignition 42 per cent. of water. Amorphous. The hardness of samoite is 4, structure laminated, and loses by ignition 30 per cent. The hardness of the others is 1—2. Halloysite loses on ignition 16 per cent. of water; collyrite, 33.5.

Cimolite, $2Al^2O^3$, $9SiO^2 + 3HO$; *Kaolinite*, $(\frac{1}{4}(HO)^3 + \frac{3}{4}Al^2O^3)^2 3SiO^2$, are very soft and earthy, and but little affected by acids; lose on ignition from 12 to 16 per cent. of water. Nearly related to these minerals are the various varieties of common clay, some varieties of lithomarge (with 14 per cent. of water), and bole with 24—26 per cent. of water; the clays become plastic with water, the latter two not.

Compare also *Lazulite*, *Svanbergite*, *Pyrophyllite*, *Agalmatolite*, which yield water in matrass, but only a very little. Compare also *Ripidolite*.

Section 2. Giving little or no water in a matrass.

Alumian, Al^2O^3, SO^3. Before the blow-pipe on coal, sulphur separates.

Lazulite, PO^5, $Al^2O^3 + MgO$, HO. Gives the reaction of phosphoric acid, § 74. Heated, loses its blue color and becomes white. Not affected by acids.

Svanbergite, PO^5, SO^3, Al^2O^3, CaO, etc. On coal, sulphur separates; color, yellow, yellowish-brown.

Willemite, $2ZnO$, SiO^2. With solution of cobalt (§ 44) becomes blue, and green in spots. Gelatinizes with hydrochloric acid. See § 215.

Myelin, Al^2O^3, SiO^2; *Agalmatolite*, SiO^2, Al^2O^3, KO, HO; *Pyrophyllite*, $(\frac{1}{5}(3HO) + \frac{4}{5}Al^2O^3)$

CLASS III. — Infusible, or fusibility above 5.
Division 1 — (*continued*).

$3SiO^2 + \frac{1}{5}HO$. Are very soft. Hardness, 1—2. Pyrophyllite is foliated like talc; before the Blp. swells up and spreads out into fan-like shapes, increasing to about 20 times its former bulk. The others do not change before the Blp. Myelin is partially decomposed by hydrochloric acid; agalmatolite not affected.

Muscovite, $3((KO)^3 (Al^2O^3))^2 3SiO^2 + 2((KO)^3 (Al^2O^3))3SiO^2$. Cleavage eminent in one direction; folia elastic. Does not swell perceptibly before the Blp.; fusible in very thin laminæ. Not affected by acids. Hardness, 2.5.

Disterrite (variety of *Seybertite*). Cleavable in one direction. Hardness, 4—5. Decomposed by concentrated sulphuric acid.

Andalusite, Al^2O^3, SiO^2; *Cyanite*, Al^2O^3, SiO^2, are but little affected by acids. Cyanite occurs generally in bladed crystallizations; hardness, 6—7. Hardness of andalusite, 7.5; but variety chiastolite varies in hardness from 3 to 7.5.

Sillimanite, *Wörthite*, *Monrolite* (vars. of *Fibrolite*), are closely related.

Topaz, Al^2O^3, SiO^2; *Rubellite* [Tourmaline], SiO^2, BO^3, Al^2O^3, MnO, LiO, KO. Not affected by acids. Not completely soluble in S. Ph., the glass becomes opalescent on cooling. Topaz on being ignited remains transparent and does not swell; tourmaline becomes white and swells. Topaz is cleavable in one direction; hardness, 8. Tourmaline is not cleavable; hardness, 6.5.

Corundum [Sapphire], Al^2O^3; *Chrysoberyl*, BeO, Al^2O^3. Not affected by acids. When pulverized, slowly but completely soluble in S. Ph.; the glass does not opalesce on cooling. Hardness of chrysoberyl, 8.5; of corundum, 9. Color of the former usually green; of the latter, blue, red, yellow, brown. Compare *Spinel.*

(Some varieties of *Leucite* assume a blue color with solution of cobalt, but its hardness is not over 6. *Cassiterite*, also, in fine powder, takes a blue color, also green. Gives with cyanide of potassium globules of tin. *Quartz* also takes a pale blue color with a reddish tinge.)

Division 2. Moistened with solution of cobalt and ignited, assume a green color.

It is sufficient to heat to redness. The minerals of this division give a coating of oxide of zinc on charcoal, § 25.

Smithsonite, see § 213.

Hydrozincite [Zinc Bloom], ZnO,CO^2 + 2(ZnO, HO). Dissolves readily in hydrochloric acid with effervescence; the solution gives with ammonia a white precipitate, soluble in an excess of the reagent. Yields water in a matrass.

Willemite, see § 215; *Calamine*, see § 214. Gelatinize with hydrochloric acid. Calamine yields water, willemite not. With solution of cobalt, assume a green color only in spots.

(See *Sphalerite* and *Goslarite*, also *Cassiterite.*)

Division 3. After ignition have an alkaline reaction, and change to blue the color of a moistened red litmus paper.

Brucite, MgO, HO; *Hydrodolomite*, (CaO, MgO)

CLASS III.—Infusible, or fusibility above 5.
Division 3—(*continued*).

$CO^2 + \frac{1}{3}HO$; *Hydromagnesite*, $MgO,HO + 3(MgO,CO^2 + HO)$. Yield much water in a matrass, unlike the other minerals of this division. Brucite dissolves in hydrochloric acid without effervescence, hydromagnesite with effervescence. The concentrated hydrochloric acid solution of the hydromagnesite is not precipitated by sulphuric acid, while the latter yields a heavy precipitate. *Predazzite*, $2CaO,CO^2 + MgO, HO$, and *Pencatite*, $CaO,CO^2 + MgO, HO$, are similar in behavior to the hydrodolomite. *Pyrochroite*, MnO,HO, is similar in reactions to brucite, but boiled with concentrated phosphoric acid, gives, on addition of nitric acid, a violet-red solution. *Lancasterite*, is a mixture of brucite and hydromagnesite. *Nemalite* is a fibrous variety of brucite, of silky lustre.

Calcite, $CaO. CO^2$; *Aragonite*, $CaO. CO^2$. Dissolve readily and with effervescence in dilute cold hydrochloric acid; the concentrated (but not the dilute) solution gives a precipitate with sulphuric acid. Aragonite falls to powder before the Blp., calcite not. (See *Strontianite*.)

Dolomite, $MgO. CO^2 + CaO. CO^2$; *Magnesite*, $MgO. CO^2$. Do not, or but slightly, effervesce with cold dilute hydrochloric acid, but dissolve readily on application of heat. The concentrated solution of the former gives a precipitate with sulphuric acid, that of the latter not.

A similar behavior shows the *Breunnerite*, (MgO.

FeO. MnO), CO^2 (ferriferous magnesite,) which on ignition becomes black and slightly magnetic; and some varieties of *Siderite*, see § 163, and *Diallogite*, see § 185.

Strontianite, $SrO.CO^2$; *Barytocalcite*, $BaO.CO^2 + CaO.CO^2$. Dissolve with effervescence in dilute hydrochloric acid; the solution, even if largely diluted with water, gives a precipitate with sulphuric acid. Strontianite colors the flame red, § 34; barytocalcite, yellowish-green, § 35.

(See also *Yttrocerite.*)

Division 4. Completely soluble, or nearly so, in hydrochloric or nitric acid, without gelatinizing or leaving a perceptible residue of silica.

Cervantite, $SbO^3 + SbO^5$, before the Blp., on coal, infusible, but with soda easily reduced to metallic antimony. Color, yellowish.

Stibiconite, SbO^5, and *Volgerite*, $SbO^5 + 5HO$, are similar. They yield water in the matrass 5 per cent., and 15 per cent.

Siderite, see § 163; *Breunnerite*, see preceding division; *Diallogite*, see § 185; *Zaratite*, see § 195. Dissolve in heated hydrochloric acid with effervescence.

Mesitite, $2MgO. CO^2 + FeO. CO^2$. Blackens and becomes magnetic before the Blp. Slightly acted upon in the cold by acids; but, if powdered, dissolves readily with effervescence in hot hydrochloric acid. Color, yellowish-white to brown. Streak, nearly white.

Ankerite, $CaO. CO^2 + (MgO, FeO, MnO)CO^2$, is similar to the last. If dissolved in aqua regia, the iron precipitated by ammonia, a heavy pre-

CLASS III. — Infusible, or fusibility above 5.
Division 4—(*continued*).

cipitate will be formed on addition of oxalate of ammonia.

Hydrotalcite, $6MgO.HO + Al^2O^3, 3HO + 6HO$. Yields water in matrass. Does not become magnetic in the reduction flame. In powder effervesces with hydrochloric acid and dissolves completely. If the solution is neutralized with carbonate of soda and filtered, oxalate of ammonia gives no precipitate in the filtrate, but phosphate of soda and ammonia do.

Parisite, $(CeO, LaO, DiO)CO^2 + \frac{1}{3}(Ca, Ce) F$, is slowly soluble in hydrochloric acid with effervescence. The solution, not too acid, gives a white precipitate with oxalic acid which becomes brick-red by ignition.

Limonite, see § 155; *Göthite,* Fe^2O^3, HO. Become black and magnetic in the reduction flame. Dissolve in hydrochloric acid without effervescence. Göthite occurs crystallized, and cleaves distinctly in one direction. Color, hyacinth-red, also brown and blackish-brown. Loses 10 per cent. on ignition. Limonite loses 14.5 per cent. Streak of both, ochre-yellow.

Turgite, $2Fe^2O^3, HO$, has a brownish-red powder, and loses by ignition 5.7 per cent. of water.

(See also *Hematite,* which in some varieties is without metallic lustre; readily distinguished by red streak.)

Blende, see § 212 (var. *Marmatite,* $FeS + 3ZnS$); *Greenockite,* CdS. Dissolve in hydrochloric acid with evolution of sulphuretted hydrogen. Give the sulphur reaction, § 107. Greenockite gives on charcoal a coating of oxide of cadmium, § 24, the others of oxide of zinc, § 25. Marmatite gives, after calcination with the fluxes, the reactions of iron.

Wad, see § 184; *Zincite*, see § 211.

Asbolite (var. of *Wad*), see § 132. Some varieties are fusible.

Uraninite, UO, U^2O^3; *Zippeite*, $(U^2O^3 3CuO)^3 2SO^3 + 8HO$. Give with the fluxes the reactions of sesquioxide of uranium [Table II]. Give with nitric acid a yellow solution in which ammonia produces a sulphur-yellow precipitate. Uraninite is black; zippeite, yellow. Sp. gr. of uraninite, 6.5.

Turquois, $(Al^2O^3)^2PO^5 + 5HO$. Color, sky-blue and green. Gives the copper reaction, § 74. Yields much water in a matrass. Sp. gr., 2.6—2.8.

Apatite, $(3CaO. PO^5) + \frac{1}{3}Ca$ (Cl, F). Gives the phosphoric acid reaction, § 94. Fusibility, 5. Soluble in nitric acid. If the solution is not too acid, a precipitate of phosphate of lead is formed with the acetate, and of oxalate of lime with oxalate of ammonia. Sp. gr., 3.2.

Monazite, PO^5, CeO, LaO, ThO, DiO. Infusible. Gives the phosphoric acid reaction, § 94. Soluble in hydrochloric acid. Minute tabular crystals of reddish-brown color. Sp. gr., 4.9—5.2.

Childrenite, $2(FeO, MnO)^4PO^5 + (Al^2O^3)^2, PO^5 + 15 HO$. Gives the phosphoric acid reaction, § 94. With the fluxes gives the reaction of iron and

CLASS III.—Infusible, or fusibility above 5.
Division 4—(*continued*).

manganese. In hydrochloric acid soluble with difficulty. Yields much water. Sp. gr., 3.2.

Polycrase, TiO^2, CbO^3, Zr^2O^3, Fe^2O^3, Ce^2O^3, UO, etc. Decrepitates, but infusible. Color, black. On fusing the pulverized mineral with hydrate of potassa, dissolving the fused mass in water, neutralizing the filtrate with H. Cl., a precipitate is formed, which, boiled with an excess of concentrated hydrochloric acid and tin foil, gives a cloudy blue solution, which filters clear and blue after the addition of a little water. This solution colors turmeric paper orange-yellow. Sp. gr., 5.

Fluocerite, $CeF + Ce^2F^3$. Gives the reactions of fluorine, § 75, and of sesquioxide of cerium, Table II. *Yttrocerite*, CaF, CeF, YF., behaves similarly.

Division 5. With hydrochloric acid gelatinize, or decompose with separation of silica without gelatinizing.

Section 1. Giving water in a matrass.

Dioptase, CuO, $SiO^2 + HO$; *Chrysocolla*, see § 149. Behave alike before the Blp; the former gelatinizes with acids, the latter not.

Xonaltite, CaO, $SiO^2 + \frac{1}{4}HO$. Massive; very hard; white to gray; yields water; infusible (?); decomposed by hydrochloric acid, in which solution of oxalate of ammonia gives a heavy precipitate, but ammonia none.

Thorite, ThO, $SiO^2 + 1\frac{1}{3}HO$; *Cerite* (CeO, LaO,

DiO)^2 SiO^2 + HO. Gelatinize with hydrochloric acid. The solution of cerite, not too acid, gives with oxalate of ammonia a white precipitate, which becomes brick-red if ignited on platinum. Color of thorite is black; streak, dark-brown; hardness, 4.5—5; of cerite, brown to red, passing into gray; streak, white; hardness, 5.5. Their sp. gr. is 4.7—5.

Chloropal, $(3FeO, Fe^2O^3)3SiO^2 + 4\frac{1}{2}HO$; *Wolchonskoite,* $MgO, Al^2O^3, Cr^2O^3, Fe^2O^3, SiO^2, HO$, and *Genthite,* $[\frac{1}{3}HO + \frac{2}{3}(NiO, MgO)]^2SiO^2 + \frac{4}{3}HO$. Amorphous, with resinous lustre. Wolchonskoite is dark sea-green, and gives with borax an emerald-green bead, which continues when cold. The others are yellowish-green. Chloropal gives a green bead, which fades on cooling, and genthite a violet bead in O. Fl., becoming gray in R. Fl. If the mineral is powdered and moistened with potassa, the chloropal becomes black without boiling; genthite turns brown after boiling until concentrated, and wolchonskoite is not changed. Genthite gives off water in the closed tube and blackens.

Gillingite, $[(CaO, MgO, FeO)^3Fe^2O^3]^2\ 3SiO^2 + 6HO$; *Xylotile,* $Fe^2O^3, 3SiO^2 + 3MgO, 2SiO^2 + 5HO$. Become magnetic by ignition. Readily decomposed by hydrochloric acid. Gillingite is black; amorphous. Xylotile is light or dark-brown, of fibrous, woody structure.

Sepiolite, $(\frac{2}{3}MgO + \frac{1}{3}HO)\ SiO^2 + \frac{1}{3}HO$. Gelatinizes with hydrochloric acid; very light; sp. gr., 1.5; absorbs water with great avidity; gives the magnesia reaction with solution of cobalt, § 44.

17*

CLASS III. — Infusible, or fusibility above 5.

Division 5 — (*continued*).

Before the Blp., turns white and shrinks. Forms a jelly-like mass with hydrochloric acid.

Bastite or *Schiller-Spar*, $3([MgO. FeO], SiO^2) + 2(MgO, 2HO)$; *Chrysotile*, $3MgO, 2SiO^2 + MgO, 3HO$. Possess a metallic pearly lustre; the former is massive, cleavable; the latter fibrous. By ignition schiller-spar becomes brown; chrysotile, white. Both are decomposed by hydrochloric acid, or more readily by sulphuric acid, without gelatinizing. *Metaxite* is greenish-white, similar to chrysotile.

Cerolite, $(\frac{1}{2}HO + \frac{1}{2}MgO)SiO^2 + \frac{1}{2}HO$. Amorphous. H = 2—2.5. G = 2.3. Color, greenish, yellowish, reddish. Before the Blp., blackens but does not fuse. Ignited with solution of cobalt, a pale flesh-red color.

Serpentine, $(\frac{3}{4}MgO + \frac{1}{4}HO)^2SiO^2 + \frac{1}{2}HO$. Decomposed by concentrated hydrochloric acid without gelatinizing. Usually massive and compact; hardness, 3—4; loss by ignition, 12 to 13 per cent. Of similar composition, and showing a similar behavior, are the following minerals, which, however, possess crystalline structure and cleavage: *Picrophyll*—Fibrous; greenish-gray; hardness, 2.5; loss by ignition, 10½ per cent. *Picrosmine*—Greenish-white, dark-green, gray; hardness, 2.7; loss by ignition, 9 per cent. *Marmolite*—Greenish and bluish-white; hardness, 2.5—3; loss by ignition, 15.7 per cent. *Kæmmererite*—Hardness, 1.5—2; loss by ignition, 13 per cent.

(See also *Chlorite* and *Ripidolite*, which are with difficulty decomposed by concentrated hydrochloric acid.)

Antigorite, $3(MgO.FeO),2SiO^2+MgO, HO$; *Monradite*, $MgO, FeO, SiO^2+\frac{1}{4}HO$; *Neolite*, $MgO, SiO^2 + \frac{1}{6}HO[+\frac{1}{9}(Al^2O^3, 3HO)]$. Decomposable by concentrated hydrochloric acid without gelatinizing. Loss by ignition, 4 to 6 per cent. Antigorite occurs in foliated masses; hardness, 2.5. Monradite, hardness, 6. Neolite in silky fibres or massive; hardness, 1.

(See, also, some varieties of *Seybertite*. Hardness, 4—5. Lustre, pearly submetallic. Color, reddish-brown, copper-red.)

Section 2. Giving only traces or no water in a matrass.

Gadolinite, $SiO^2(YO, FeO, CeO, BeO)^2$; *Gehlenite*, $(\frac{1}{2}(MgO, FeO, CaO)^3 + \frac{1}{2}(Fe^2O^3, Al^2O^3))^2 SiO^2$. Gelatinize with hydrochloric acid. Gadolinite swells before the Blp. into cauliflower-like masses, and sometimes exhibits a vivid glow; thin splinters fusible on the edges; color, black to blackish-green; hardness, 6.5—7. G = 4—4.3. Gehlenite is also fusible in very thin splinters; color, gray to grayish-white; hardness, 5.5—6. G = 3.

Chrysolite, $(MgO, FeO)^2 SiO^2$. H = 7. *Chondrodite*, $8MgO, 3SiO^2$, with part of the oxygen replaced by fluorine. H = 6.5. Gelatinize with hydrochloric acid. Color of the former, green; of the latter mostly white, yellow, or brown. Chondrodite gives the fluorine reaction, § 76. *Monticellite* is isomorphous with chrysolite.

CLASS III.—Infusible, or fusibility above 5.

Division 5 — (*continued*).

H=5—5.5. Color, yellowish, greenish-gray, and white.

Forsterite, $2MgO, SiO^2$. Cleavage distinct in one direction. Color, white, grayish, yellow, wax-yellow. Before the Blp., unaltered and infusible. Decomposed by H. Cl. with separation of gelatinous silica. Var. *Boltonite* gives traces of moisture in matrass and becomes colorless. (See also *Seybertite*.)

Leucite, $KO, SiO^2 + Al^2O^3, 3SiO^2$. Decomposed by hydrochloric acid, the silica separating as a fine powder; some varieties become blue with solution of cobalt; occurs usually in trapezohedrons. Color, grayish or white. H = 5.5. G = 2.5.

Division 6. Not belonging to either of the preceding divisions.

The remaining minerals, which cannot be classed under any of the preceding divisions, may be divided according to their hardness in two sections.

Section 1. Hardness below 7.

Biotite [Hexagonal Mica], $\frac{1}{2}(Al^2O^3. Fe^2O^3)^2, SiO^2 + \frac{1}{2}(MgO, KO)^3, 3SiO^2$; *Muscovite* [Oblique Mica], $3(3RO, R^2O^3)^2\ 3SiO^2 + 2(3RO, {}^2RO^3)3\ SiO^2$, [$RO = KO, NaO, MgO, R^2O^3 = Fe^2O^3, Al^2O^3$]; *Talc*, $(\frac{1}{5}HO + \frac{4}{5}MgO)SiO^2$. Give little or no water in a matrass. Talc loses at most 5 per cent. Cleavage eminent in one direction. Hardness of biotite, 2.5—3; of muscovite, 2—2.5;

of talc, 1—1.5. Biotite is decomposed by concentrated sulphuric acid, the others not. The laminæ of biotite and muscovite are elastic; of talc not. Soapstone or *Steatite* is a massive, usually compact, variety of talc; very greasy to the feel, or like soap. (See also *Pyrophyllite.*) *Margarodite* and *Phlogopite* are decomposed by sulphuric acid. *Margarite*, with pearly lustre, and *Œllacherite*, with $5\frac{1}{2}$ per cent. baryta, are nearly related to the muscovite.

Prochlorite, $(\frac{4}{7}(MgO, FeO)^3 + \frac{3}{7}Al^2O^3)SiO^2 + \frac{4}{3}HO$; *Ripidolite*, $5MgO, Al^2O^3, 3SiO^2, 4HO$. Lose by ignition 12 per cent. of water. Cleavage eminent in one direction, laminæ not elastic (both often massive-granular). Hardness of prochlorite, 2; of ripidolite, 2—2.5. Decomposed by concentrated hydrochloric acid, with continued boiling, more readily by sulphuric acid. Ripidolite fuses with difficulty (5.5) to a grayish-black glass; prochlorite becomes black and slightly magnetic. Ripidolite gives, with borax, a clear glass, colored by iron and sometimes chromium. Ripidolite is monoclinic; prochlorite, hexagonal.

Leuchtenbergite, H = 2.5. Colorless, white, yellowish-white. Before the Blp. exfoliates and fuses with difficulty on the thin edges, becoming white and opaque. *Penninite*, H = 2—25; 3 often on edges. Color, green, red, and white. With fluxes, all varieties give the reaction for iron, and many for chromium. Completely decomposed by sulphuric acid. *Chloritoid* is not perceptibly acted upon by hydrochloric acid, but completely decomposed by sulphuric acid. H = 5—6. Loss by ignition, $7\frac{1}{2}$ per cent.

CLASS III.—Infusible, or fusibility above 5.
Division 6—(*continued*).

Cerolite [compare Div. 5, Sect. 1]. Amorphous. Yellowish-white. H = 2—3. Loss by ignition, 30 per cent. Mostly decomposed by hydrochloric acid.

Beauxite, (Al^2O^3, Fe^2O^3)2HO. Amorphous. Often oölitic, grayish, reddish-brown, and red. H = 3. G = 2.5. Loss by ignition, 20 per cent. Only slightly attacked by hydrochloric acid, but completely dissolved by concentrated phosphoric acid. (Compare *Argillite.*)

Wolchonskoite [see Div. 5]. Amorphous. Color, dark-green. Boiled with phosphoric acid, it gives an emerald-green solution, which, if diluted with water, retains its color, but gelatinous silica separates out. *Chromite* also gives the chromium reaction, but its color is black, and streak yellowish-brown.

Warwickite, (MgO, FeO, TiO^2, BO^3.) Its powder is decomposed by sulphuric acid; evaporated to dryness and moistened with alcohol, it gives the green flame. If this mass is boiled with hydrochloric acid and tin foil, and concentrated, the solution is violet, or, diluted with water, rose-red.

Enstatite, MgO, SiO^2; *Anthophyllite*, ($\frac{1}{4}$FeO + $\frac{3}{4}$ MgO)SiO^2. Cleavage of enstatite very perfect in one direction; anthophyllite cleaves in two directions, under 124° 30′. The former is of clove-brown or pinchbeck-brown color, with a pearly-metallic lustre; the lustre of anthophyllite is much less perfect. Hardness, 5—5.5. *Hypersthene*, (MgO, FeO)SiO^2, is closely related, and cleaves at 86½°. H = 5.5.

Tungstite, WO^3. Boiled with phosphoric acid it gives a bluish solution, which, shaken while warm with iron filings and a little water, becomes dark-blue. Occurs in soft, earthy, yellow masses.

Scheelite, CaO, WO^3. Fusibility, 5; hardness, 4.5—5. The pulverized mineral, on being boiled with nitric acid, leaves a lemon-yellow residue of tungstic acid. Gives the reactions of tungstic acid [Table II].

Cassiterite, see § 209.

Octahedrite, *Rutile* (both tetragonal, with adamantine lustre), and *Brookite* (orthorhombic), TiO^2. Give the reactions of titanic acid [Table II]. On fusing the pulverized minerals with hydrate of potassa, dissolving the fused mass in hydrochloric acid and boiling the solution with metallic tin, it assumes a violet color, which turns to red on addition of water. Color of octahedrite, various shades of brown, passing into indigo-blue; of rutile, mostly brownish-red or red, sometimes yellowish or black; of brookite, hair-brown, yellowish, or reddish (variety *Arkansite* is iron-black). Hardness of octahedrite, 5.5—6; of rutile, 6—6.5; of brookite, 5.5—6.

Euxenite and *Aeschynite*, CeO, LaO, YO, FeO, TiO^2, TaO^5, CbO^5; *Pyrochlore*, $(CaO, CeO)^2$, CbO^5.(?) Treated like the preceding with potassa, etc., the solution on reaching a certain degree of concentration assumes a fine blue color, on

CLASS III. — Infusible, or fusibility above 5.
Division 6 — (*continued*).

addition of water, which changes in the air to olive-green and gradually disappears. Aeschynite swells before the Blp. and turns yellow or brownish. The color is black; the powder, light-brown. Euxenite unaltered before the Blp. Color, brownish-black. Powder, reddish-brown. They have a metallic, greasy lustre. Pyrochlore is distinguished by its octahedral form. Color, brownish-red, and powder, light-yellow.

Opal, SiO^2. Amorphous. Before the Blp. yields water and becomes opaque; fuses with carbonate of soda to a clear bead, with effervescence. Infusible. Boiled with hydrate of potassa, it dissolves completely or to a great extent; the solution gives a gelatinous precipitate with chloride of ammonium. Hardness, 6—6.5. G=2—2.3.

Xenotime, $3YO, PO^5$. Color, various shades of brown or flesh-red. Hardness, 5. G = 4.4. Gives the phosphoric acid reaction, § 94. Infusible. With salt of phosphorus dissolves with great difficulty to a colorless glass.

(See also *Childrenite* and *Orthoclase*.)

Section 2. Hardness, 7, or above.

(See *Cassiterite*, *Rutile*, and *Opal* of the preceding section, whose hardness sometimes approaches 7.)

Quartz, SiO^2. The various varieties of quartz, as *Rock-crystal*, *Amethyst*, *Hornstone*, *Flint*, *Chalcedony*, etc., are infusible and unalterable before the Blp., and fuse with carbonate of soda to a transparent bead, with effervescence. Hardness, 7. G = 2.6.

Iolite, $2(FeO, MgO)SiO^2 + 2Al^2O^3, 3SiO^2$; *Staurolite*, $(\frac{1}{5}(3FeO)+\frac{4}{5}Al^2O^3)^4 3SiO^2$. Do not fuse to a transparent glass with carbonate of soda. Fusibility of iolite, 5—5.5; color, blue, grayish. G = 2.6. Staurolite is infusible; color, brownish-red, brown; crystals often cruciform. G = 3.6.

Beryl, $(\frac{1}{2}(3BeO) + \frac{1}{2}Al^2O^3)3SiO^2$; *Euclase*, $(\frac{1}{6}(3HO) + \frac{2}{6}(3BeO) + \frac{3}{6}Al^2O^3) SiO^2$; *Phenacite*, $2BeO, SiO^2$; *Zircon*, ZrO, SiO^2. Hardness, 7.5. Beryl and euclase turn milk-white with strong heat and become rounded on the edges; beryl crystallizes in hexagonal prisms, and possesses pretty distinct basal cleavage; color usually pale-green or emerald-green. Euclase crystallizes in clinorhombic prisms, and possesses distinct cleavage in two directions at right angles to each other; color, pale mountain-green, passing into blue and white. Phenacite and zircon do not change before the blow-pipe, excepting that zircon becomes colorless; color, red, yellow, or colorless; zircon, sometimes brown or gray; phenacite is a little harder (8) than zircon. G. of zircon, 4.4; of the others, 2.7—3.

Topaz, Al^2O^3, SiO^2. H = 8. G = 3.5. Orthorhombic. Before the Blp. infusible. Some varieties take a wine-yellow or pink tinge when heated. Fused in an open tube with salt of phosphorus gives the fluorine reaction. With solution of cobalt the pulverized mineral gives a fine blue on heating. Very slightly attacked by sulphuric acid.

Ouvarovite [Lime-Chrome-Garnet], $(\frac{1}{2}(3CaO) +$

CLASS III.—Infusible, or fusibility above 5.
Division 6—(*continued*).

$\frac{1}{2}Cr^2O^3)^2$ $3SiO^2$, Cr^2O^3, SiO^2. Emerald-green. Infusible, but by ignition becomes blackish-green, and on cooling again, emerald-green. Hardness, 7.5—8. G = 3.5. Gives with fluxes the chromium reactions [Table II].

Spinel, MgO, Al^2O^3; *Pleonaste,* (MgO. FeO), Al^2O^3; *Gahnite,* $ZnO.Al^2O^3$. Hardness, 7.5—8. Occur almost exclusively in octahedral crystals. Spinel and pleonaste, when pulverized, are soluble in salt of phosphorus; color of spinel, red, blue, brownish; of pleonaste black. Gahnite is almost insoluble in salt of phosphorus and borax; color, dark-green or black. *Kreittonite* is a black spinel containing zinc and iron, slightly magnetic before ignition; so also *Dysluite.*

Diamond, C. Characterized by its hardness, which surpasses that of corundum. H = 10. G = 3.5—3.6.

THE OXIDIZED MINERALS, ARRANGED ACCORDING TO THEIR FUSIBILITY AND BEHAVIOR WITH CARBONATE OF SODA.

(From *Plattner's Blow-pipe Analysis.*)

a. Minerals fusible to a bead.

a. With soda yield a fluid bead:

Aconite,	Elaeolite,*	Oligoclase,
Allanite,*	Eudialyte,	Pyrasmolite,
Axinite,*	Garnet,	Sassolite,*
Boracite,*	Helvite,	Scapolite,*
Borax,*	Hydroboracite,*	Sodalite (Greenland),
Botryolite,*	Ilvaite,	Spodumene,
Crocidolite,	Labradorite,	Talc, black,
Cronstedtite,	Lapis Lazuli,	The Zeolites.*
Datolite,*	Mica, from primitive limestone,	

b. With a little soda a bead, with more a slaggy mass:

Amblygonite,	Okenite,	Rhodonite,*
Fluorite,	Orthite,*	Sodalite,
Garnet, manganiferous,	Pectolite,	Sordavalite,
Manganese, black silicate (hydrous tephroite?),	Pyrorthite,	Vesuvianite.*

c. With soda only a slag:

Amphodelite,	Iron, phosphates of sesquioxide,	Saponite,
Autunite,	Pharmacolite,	Scarodite,
Brevicite,	Pharmacosiderite,	Tourmaline, potash, (black),
Fahlunite,	Polyhalite,	Triphylite,
Haüynite,	Pyrargillite,	Wolframite.
Heterosite,	Pyrope,	

d. Sink with soda into the charcoal:

Celestite, Witherite.

* Denotes that the mineral fuses with intumescence, effervescence, etc.

e. Fuse with soda at first more or less perfectly to a clear mass, but are decomposed by a sufficient quantity of soda, and leave behind an infusible crust, while the soda salt sinks into the coal:

Anhydrite,	Gay-Lussite,	Gypsum,
Cryolite,	Glauberite,	Polyhalite.

f. Yield with soda reguline metal: Minerals consisting of reducible metallic oxides and their reducible salts.

b. Minerals which fuse only on the edges.

a. With soda yield a fluid bead:

Albite,	Nephelite,	Steatite,
Anorthite,	Orthoclase,	Titanite,*
Emerald (beryl),	Petalite,	Turquois.
Euclase,*	Sodalite (Vesuvius),	

b. With little soda a fluid bead; more, a slaggy mass:

Enstatite, Epidote,* Hypersthene, Wollastonite, Zoisite.*

c. Yield with soda only a slag:

Carpholite,	Pimelite,	Scheelite,
Dichroite (iolite) blue,	Pinite,	Serpentine,
Lazulite,*	Plumbogummite,*	Tourmaline,* soda,
Mica,* from granite,	Pyrochlore,	(green).

d. With soda goes into the coal:

Barite.

e. Fuse or only swell up with soda, but are decomposed by a sufficient amount of soda, leaving an infusible crust, while the soda salt sinks into the coal:

Apatite (swells up), Barite (calciferous), fuses.

* Denotes that the mineral fuses with intumescence, effervescence, etc.

c. Infusible Minerals.

a. Give with soda a fluid bead:

Agalmatolite,
Dioptase,
Fire-clay,
Hisingerite,
Leucite,
Pyrophyllite,*
Quartz,
Rutile,
Sideroschisolite,
Wolchonskite.

b. With little soda, a bead; with more, a slaggy mass.

Cerite,
Chrysolite,
Gadolinite,*
Phenacite,
Picrosomine,
Talc,
Tourmaline* (lithia).

c. Yield, with soda, only a slag:

Aeschynite,*
Allophane,
Aluminite,
Alunite,
Alunogen,*
Andalusite,
Brucite,
Calamine (a zinc coat),
Cassiterite, with much soda metallic tin,
Chloritoid,
Chromite,
Chrome Ochre,
Chrysoberyl,
Cyanite,
Diaspore,
Fluocerine,
Gahnite (a zinc coat),
Gehlenite,
Gibbsite,
Iron, sesquioxide and its sulphates,
Manganese, oxides,
Oerstedite,
Ouvarovite,
Polymignite,
Spinel,
Staurolite,
Tantalite,
Thorite,
Titanic Iron,
Topaz,
Wörthite,
Xenotime,
Yttrocerite,
Yttrotantalite,
Zircon.

d. Fuse or only swell up with soda, but are decomposed by a sufficient amount, and the soda sinks into the coal, leaving an infusible crust:

Alum (kalinite),
Aragonite,
Barytocalcite,
Calcite,
Dolomite,
Epsomite,
Magnesite,
Wavellite.*

e. Sinks with the soda into the coal:

Strontianite.*

* Denotes that the mineral fuses with intumescence, effervescence, etc.

CHAPTER VII.

Colored Flames, Flame Reactions, and Spectrum Analysis.

MANY substances, when brought into a colorless or non-luminous flame, color it in a remarkable manner. The colorations are, in many cases, characteristic of the elements yielding them, and furnish excellent means of detecting the latter, even in the minutest quantities, with great ease and certainty. Thus soda-salts tinge the flame yellow; potassa compounds violet; lithia-salts carmine-red; and on account of this peculiarity they may be distinguished from each other by the simplest experiments.

The Bunsen lamp, with chimney, previously described (Fig. 2), is especially adapted to such observations. The substance to be tested is brought by means of the platinum wire-loop (Fig. 3) into the zone of fusion of the gas-flame. The alkalies and alkaline earths are most remarkable in their coloring effects on the flame. If we compare together various salts of the same base, we find that they all, if volatile at the temperature of the flame, give the same color, but the color differs in intensity, being strongest with the most volatile salts, and *vice versâ*. Thus, chloride of potassium gives a deeper tinge to the flame than carbonate of potassa, and carbonate a stronger than silicate of potassa. Sometimes a non-volatile compound is made to exhibit a characteristic tint by the addition of some flux or decomposing agent. Silicates which contain but a few *per cent.* of potassa, and of themselves do not color the flame, give a coloration after heating with some pure sulphate of lime, which decomposes them, producing silicate of lime and volatile sulphate of potassa.

In mixtures of several substances which may be individually detected without difficulty, when they exist separately, it usually happens that a mixed and indecisive coloration is produced, or one substance masks all the others. Thus, in a mixture of soda and potassa-salts, only a soda flame — in one of baryta and strontia-salts, only a baryta flame is evident to the unassisted eye. We have recently learned two methods of dissecting these mixed flames so as to recognize their component colors with surprising facility and distinctness.

The first method, introduced into chemistry by Cartmell, and further developed by Bunsen and Merz, consists in observing the colored flames through colored media (stained-glass, indigo solution, etc.). These act by extinguishing the color of one metal, and thus developing that of the other. A mixture of soda and potassa, which to the eye has a pure yellow flame, when seen through a deep-blue cobalt glass or a solution of indigo, exhibits the violet tint of potassa, without any traces of the yellow soda flame. Cartmell detected lithia in the presence of soda and potash, by comparing the mixed color of the flames of those bases with that of the flame of pure potash, when both are viewed through an indigo solution. Bunsen has found that the discrimination of these bases in presence of each other is more easily effected by observing the *succession of changes of color*, which the mixed flame produced by these substances experiences when the rays reach the eye after passing through gradually thicker layers of an indigo solution.

The apparatus for these observations is simple, viz.:

1. A hollow prism, made of plate glass, Fig. 9, whose principal section forms a triangle, with two sides of 150 millimètres and one of 35 millimètres long. The solution with which it is filled is prepared by dissolving 1 part of indigo in 8 parts of fuming oil of vitriol, adding 1500 to 2000 parts of water and filtering.

In the following experiments the prism was moved horizontally before the eye, so that the rays of the flame always passed through gradually thicker layers of the medium. The alkaline substances, brought singly into the melting-space, exhibited the following changes:

a. Chemically pure CaCl. produces a yellow flame, which, even with very thin layers of the indigo solution, passed through a tinge of violet into the original blue lamp flame.

b. Chemically pure NaCl., the same.

c. Chemically pure KO,CO_2 *or* KCl. appears of a sky-blue, then violet, and at last of an intense crimson-red, even when seen through the thickest layers of solution. Admixtures of soda or lime do not hinder the reaction.

d. Chemically pure LiO,CO_2 *or* LiCl. gives a carmine-red flame, which, with increasing thickness of the medium, becomes gradually feebler, and disappears before the thickest layers pass before the eye. Lime and soda are also without influence on this reaction.

2. A blue, a violet, a red and a green glass. The blue is colored by protoxide of cobalt; the violet, by sesquioxide of manganese; the red (partly colored and partly uncolored), by suboxide of copper; and the green, by sesquioxide of iron and oxide of copper. The stained glasses found in commerce and employed for ornamenting windows, generally possess the requisite shades of color.

Merz, who has made a complete investigation of this

subject, employs with these glasses Bunsen's burner, and also a flame of pure hydrogen. The substances which he describes as giving characteristic colors to the flame of Bunsen's burner, in addition to those previously known, are nitric and chromic acids, while phosphoric and sulphuric acids give a peculiar coloration to the dark core of the flame of hydrogen.

The flame of Bunsen's burner gives three sorts of color:

a. **Border Colors.** — These are of course peculiar only to the most volatile substances. To produce them, the loop of platinum wire is to be held outside of the flame about one or two millimètres from the lower portion of the outer limit.

b. **Mantle Colors.** — Those, namely, which are seen when the substance is held in the bright blue-colored mantle which forms the outer portion of the flame.

c. **Flame Colors.** — To produce these, the loop is to be held horizontally and in the hottest part of the mantle. The hydrogen flame yields another species of color, viz., the

d. **Core Colors.** — These are produced only by sulphuric and phosphoric acids, which communicate respectively a blue and green tinge to the cold core of the hydrogen flame.

The following, according to Merz, is a list of the more commonly occurring substances which color the flame, with the color they impart:

BLUE FLAMES.

(*Consult page* 216.)

Intense blue, ...	Chloride of copper.
Pale clear blue, . . .	Lead.
Light blue, ...	Arsenic.
Greenish blue, ...	Antimony.
Blue mixed with green, . .	Bromide of copper.
Blue core color, . . .	Sulphuric acid.

GREEN FLAMES.

(*Consult page* 217.)

Bronze-green border color, . .	Nitric and Nitrous acids.
" " " " . .	Ammonium compounds.
" " " " . .	Cyanogen "
Greenish-blue border color, . .	Hydrochloric acid.
Green mantle color, . . .	Boracic acid.
Gray yellow-green border color, .	Phosphoric acid.
Yellowish-green flame color, . .	Barium compounds.
Dark green, ...	Iron wire.
Full green, ...	Copper.
Intense emerald green, . .	Iodide of copper.
Emerald green, mixed with blue, .	Bromide of copper.
Pale green, ...	Phosphoric acid.
Intense whitish green, . . .	Zinc.

RED FLAMES.

(*Consult page* 218.)

Intense crimson, . . .	Strontium compounds.
Reddish purple, . . .	Calcium compounds.
Violet, ...	Potassium compounds.
Dark brownish-red border color and a rose-red mantle color, .	Chromic acid.

YELLOW FLAMES.

(*Consult page* 218.)

Yellow, ...	Sodium compounds.
Feeble brownish-yellow, . .	Water.

Blue Flames. — CuCl. gives an azure-blue zone, and CuO. NO_5 a pure green flame color. By the combined observation of both colors, Cu. may be distinguished from all other metals which give similar colors. The other flame-coloring metals, such as As., Sb., Sn., Pb., Hg., and Zn. exhibit, especially in the form of chlorides, more or less intense bluish or greenish mantle colors, which, however, cannot be advantageously used as reactions for the metals themselves.

Sulphuric acid produces a beautiful blue core color, being reduced to SO_2. The free acid gives the color when the platinum loop is held in the border of the flame, but a sulphate must be held in the middle of the flame. In the latter case, it is well to dip the test into strong HCl. or fluosilicic acid.

Green Flames. — Nitric and nitrous acids give a bronze-green border color, usually with an orange-colored border. The test is to be previously dried in the flame, and dipped into a solution of $KO, 2SO_3$, or into dilute HCl., according as we wish to test for nitric or nitrous acid. Ammoniac and cyanogen compound give the same bronze-green border, but more faintly. HCl. gives a very weak greenish-blue border color, which lasts for a very short time, and therefore does not deserve attention. The acid is, however, decomposed, and the Cl. may easily be recognized.

Boracic acid gives a beautiful green mantle color, which is so intense that the acid may be recognized in the presence of large quantities of phosphoric acid. Borates are to be decomposed with SO_3. Phosphoric acid gives a gray yellow-green border color as well as a beautiful green core color. The dry test is to be dipped into SO_3 and held in the flame in the manner already pointed out, in order to show the border color. The green core color is less sensitive, but indispensable in recognizing phosphoric acid in the presence of large quantities of boracic acid, and is produced by alternately moistening the test with a solution of fluosilicic acid, and igniting it in the hydrogen flame, until the color distinctly appears.

Ba. may be recognized by the yellowish-green flame color, which appears blue-green through the green glass. If the green disappears, and a red flame color makes its appearance, the test is to be repeatedly moistened with

HCl., and immediately introduced while wet into the hottest part of the flame. When the blue-green color is no longer seen, proceed to examine for Ca.

Red Flames. — Ca. is present when the red flame color, on evaporating the last portion of HCl., appears siskin-green through the green glass. Sr. gives in this case a weak yellow. Sr. may be recognized by the purple or rose color which is seen through the blue glass, when the test, after moistening with HCl., is evaporated to dryness in the flame.

K. gives a gray-blue mantle color, and a rose-violet flame color. These colors appear reddish-violet through the blue glass, violet through a violet glass, and blue-green through a green glass. The test is to be moistened with SO_3, and repeatedly exposed to the flame for a short time.

Chromic acid gives a dark brownish-red border color, and a rose-red mantle color. The dry test is to be moistened with concentrated SO_3, and held in the border. Chromic oxide gives no color, and is to be first oxidized to chromic acid by moistening with a solution of hypochlorite of soda and drying.

Yellow Flames. — Na. gives an orange-yellow flame color, which in very large quantities appears blue, but in small quantities is invisible through the blue glass. Through the green glass the flame appears orange-yellow, even with the smallest quantities; this glass is particularly adapted to the recognition of Na. in all its compounds. The test should be moistened with SO_3, dried and held in the hottest point of the flame.

BUNSEN'S FLAME REACTIONS.

Almost all the reactions which can be performed by means of the blow-pipe may be accomplished with greater ease and precision in the flame of the non-luminous gas-lamp. This flame, moreover, possesses several peculiarities which render it available for reactions, by which the smallest traces of many substances occurring mixed together can be detected with certainty, when the blow-pipe and even still more delicate methods fail. Only the principal reactions that can be obtained in this way are here given.

Bunsen's Gas-Lamp. — This lamp, with non-luminous flame, is represented in Fig. 2, and must be made about three times as large as the drawing. It must be furnished with a cap for closing and opening the draught-holes, so as to be able to regulate the supply of air for every dimension of the flame. The conical chimney *d d d d*, Fig. 20, must also be made of such a size that the flame burns perfectly steady. Fig. 20 represents the flame of its proper size. It is composed of the following three chief divisions:

A. **The dark cone,** *a a a a*, containing the cold unburnt gas mixed with about 62 per cent. of air.

B. **The flame mantle,** *a c a b*, formed of the burning coal-gas mixed with air.

C. **The luminous point,** *a b a*, not seen when the lamp is burning with the draught-holes open, but obtained of the size required for the reactions by closing these holes up to a certain point.

The following six points in the flame are used in the reactions:

1. **The base of the flame** lies at *α*; its temperature is comparatively very low, as here the burning gas is cooled by the upward current of cold air, and much heat is absorbed by the cold end of the metal tube. If mixtures of flame-coloring substances are held in this part of the flame, it is often possible to vaporize the most volatile constitu-

Fig. 20.

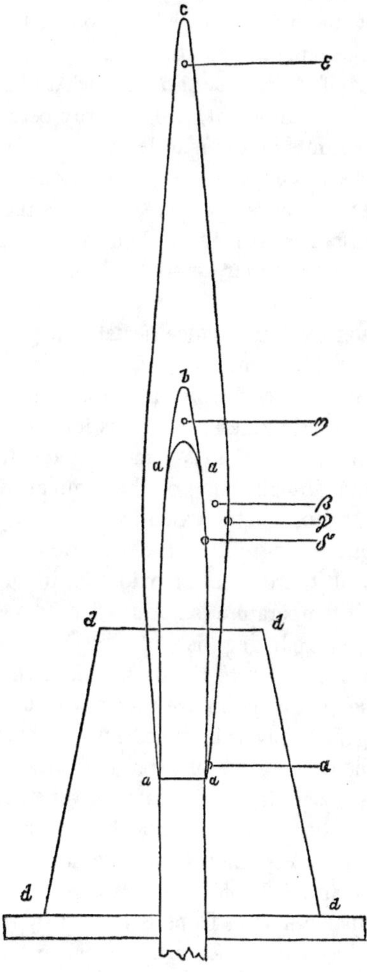

ent, and thus in the first few moments to obtain tints which cannot be observed at higher temperatures, because they then become masked by colors produced by the volatilization of the remaining substances.

2. **The zone of fusion** lies at β, somewhat above the first third of the flame in height, and midway between the inner and outer limits of the mantle at the point where the flame is thickest. This is the point in the flame which possesses the highest temperature, and it is therefore used in testing substances as regards their melting-point, their volatility, emissive power, as well as for all processes of fusion at high temperatures.

3. **The lower oxidizing flame** lies at γ, in the outer margin of the zone of fusion, and is especially suitable for the oxidation of substances dissolved in beads of fused salts.

4. **The upper oxidizing flame** at ϵ is formed by the highest point of the non-luminous flame, and acts most powerfully when the draught-holes of the lamp are wide open. This flame is suited for the oxidation of larger portions of substance, for roasting off volatile-oxidation products, and generally for all those cases of oxidation in which an excessively high temperature is not needed.

5. **The lower reducing flame** lies at δ, on the interior edge of the mantle next to the dark central zone. As the reducing gases at this point are mixed with unburnt atmospheric oxygen, many substances remain here unaltered which become deoxidized on exposure to the upper reducing flame. This point of the flame gives, therefore, very valuable reactions which cannot be obtained with the blow-pipe. It is especially available for reductions on charcoal, and in beads of fused salts.

6. **The upper reducing flame** is formed by the luminous point η, produced over the dark zone when the admission

of air is lessened by the gradual closing of the draught-holes of the lamp. If this luminous point is made too large, it will be found that a test-tube filled with cold water becomes covered with a film of lampblack: this never ought to occur. This flame contains no free oxygen, is rich in finely divided incandescent carbon, and hence it possesses far more powerful reducing powers than the lower reducing flame. It is especially available for reducing metals when it is desired to collect them in the form of films.

METHOD OF EXAMINATION IN THE VARIOUS PARTS OF THE FLAME.

A. Behavior of the Elements at High Temperatures.*

This is one of the most important reactions which can be employed for the detection and separation of substances. The possibility of producing, with the flame of the lamp alone, a temperature as high as, or higher than, that of the blow-pipe depends upon the fact that the radiating surface of the heated body be made as small as possible. The arrangement for bringing the substances into the flame must therefore be on a very small scale. The platinum wire upon which the substance is heated must scarcely exceed the thickness of a horsehair, and one decimetre in length of the wire must not weigh more than 0.034 grm. It is impossible to obtain the results hereafter detailed if a thicker wire than this is employed. Substances which act upon platinum, or which will not adhere to the moistened surface of the metal, are held in the flame

* For further information with regard to the flame reactions, the student is referred to the translation of Bunsen's paper by Professor Roscoe in the "Philosophical Magazine" for 1867.

upon a thin thread of asbestos, of which a hundred may be obtained from one splinter of the mineral. These threads

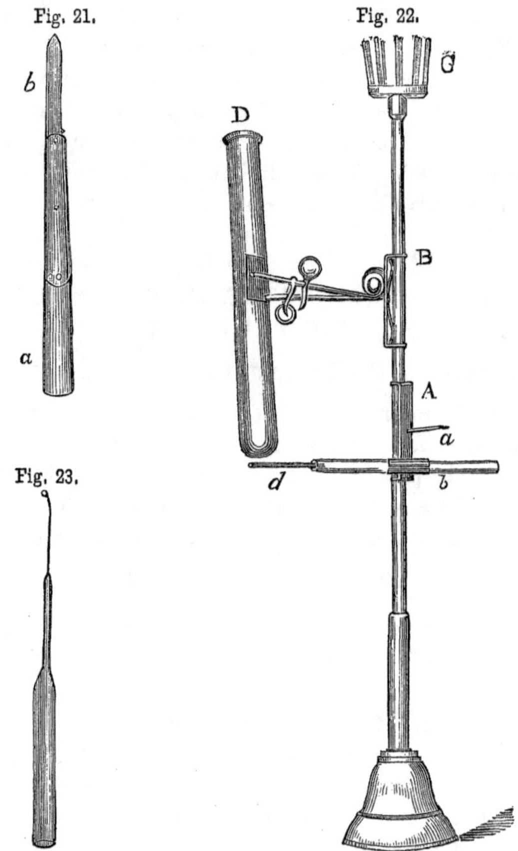

Fig. 21. Fig. 22. Fig. 23.

must not exceed in thickness one-fourth of that of an ordinary lucifer match. Decrepitating substances are ground to the finest powder on the porcelain lamp-plate with the elastic blade (*a*) of the knife (Fig. 21), and drawn up on to a moistened strip of one square centimetre of filter paper. If the paper is then burnt, being held with the platinum forceps, or, better, between two rings of fine platinum wire, the sample remains as a coherent crust, which now may without difficulty be heated in the flame.

If the substance requires to be heated in the flame for a long period, the holder (Fig. 22) is used. The arm (*a*) is fastened to the carrier (A), so fixed on the stand by a spring (as seen at B) that it can be moved both horizontally and vertically. The glass tube (Fig. 23) is held on this arm (*a*), and the fine platinum wire fused on to the tube thus held in the flame. The splinters of asbestos are stuck into the glass tube (*b*), which slips into the holder, and may then be moved with the carrier (A). The carrier (B) carries a spring-clamp for holding test-tubes which have to be heated for a considerable time in a particular part of the flame. The little turn-table (C) contains nine upright supports to hold the wire tubes (Fig. 23) employed in the experiments. By means of these arrangements a particle of the substance under examination is brought into the flame, and its behavior in the coldest and hottest parts of the flame is ascertained, the substance being examined with a lens after each change of temperature. The following six different temperatures can be obtained in the flame, and these points may be judged of by observing the tints attained by the thin platinum wire:

1. Below a red heat.
2. Commencing red heat.
3. Red heat.
4. Commencing white heat.
5. White heat.
6. Strong white heat.

It is scarcely necessary to remark that these different temperatures must not be ascertained by the glow of the substances themselves, as the luminosity of different bodies depends not only upon the temperature, but also mainly upon their specific power of emission.

The following phenomena are observed when a sample of a substance is heated:

1. Emission of Light. — The emissive power of substances is ascertained by placing them on the platinum wire in the hottest part of the flame. The sample is of weak emissive power when it is less luminous than the platinum wire; of a mean emissive power when both appear about equally luminous; and of strong emissive power when the intensity of the light which it emits is greater than that from the platinum. Most solid bodies emit a white light, others (as, for instance, erbia) colored light.

Some bodies, such as many osmium, carbon, and molybdenum compounds, volatilize and separate out finely divided solid matter, which renders the flame luminous. Gases and vapors always exhibit a smaller power of emission than fused substances, and these generally less than solid bodies. The form of the substance under examination must always be noted, as the emissive power depends upon the nature of the surface: thus compact alumina, obtained by slowly heating the hydrate, possesses only a moderate emissive power, whereas the porous oxide prepared by quick ignition of the sulphate possesses a high power of emission.

2. The melting point is determined by using the six different temperatures already mentioned. At every increase of temperature the bead is examined with the lens to see whether the volume is decreased or increased, whether bubbles are given off on melting, whether on cooling the bead is transparent, and what changes of color it undergoes during the action of the heat or on afterwards cooling.

3. **The volatility** is ascertained by allowing equally heavy beads of the substance, placed on a platinum wire, to evaporate in the zone of fusion, and observing the time, by means of a metronome, which the bead takes to volatilize. The point at which the whole of the substance is converted into vapor can be ascertained with great accuracy, often to a fraction of a second, by the sudden disappearance of the coloration of the flame. The platinum wire upon which the substance is weighed is protected from the moisture of the air by insertion in a tube (Fig. 24). If we know the weight of the tube and wire, the right weight of substance can easily be attached, either by volatilizing a portion or by fusing some more substance on to the bead, and thus making it lighter or heavier. The experiments are best made with one centigramme of substance. The position in the flame where the highest constant temperature exists can be found by moving a fine platinum wire, fixed on a stand and bent at its point at a right angle, slowly about the zone of fusion, and noting the point where it glows most intensely. The beads to be volatilized are then most carefully brought into the flame at the same distance from the point of this wire. Care must also be taken that the dimensions of the flame do not undergo change from alterations in the pressure of the gas while the experiments are going on.

Fig. 24.

4. **Flame coloration.**—Many substances which volatilize

in the flame may be detected by the peculiar kinds of light which their glowing gases emit. These colorations appear in the upper oxidizing flame when the substance causing them is placed in the upper reducing flame. Mixtures of various flame-coloring substances are tested in the lowest and coldest part of the flame; and here it is often possible to obtain for a few moments the peculiar luminosity of the most volatile of the substances unaccompanied by that of the less volatile constituents.

B. Oxidation and Reduction of Substances.

In order to recognize substances by the phenomena exhibited in their oxidation and reduction, and to obtain them in a fit state for further examination, the following methods are employed:

1. Reduction in glass tubes is especially employed for the detection of Hg., or for the separation of S., Se., P., etc., when in combination with Na. or Mg. A stock of very thin glass tubes is prepared, each 2 to 4 millims. in width and 3 centims. in length; forty of these are easily made out of one ordinary-sized test-tube, by softening the glass before the blow-pipe, and then drawing it out until the requisite size of tube is obtained. This long tube is then cut up with a diamond into pieces 6 to 8 centims. long, and each of these again divided into two over the lamp, and the closed ends neatly rounded. The sample, having been finely powdered with the knife-blade (Fig. 21, *a*) on the porcelain plate (Fig. 25),

Fig. 25.

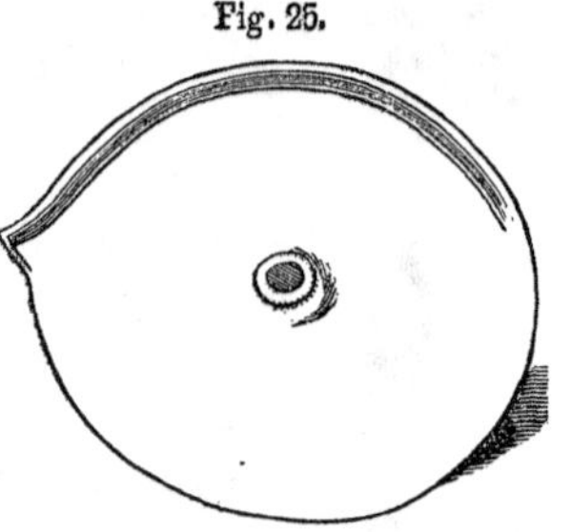

is treated in a tube either by itself, or with a mixture of carbon and soda, or with Na. or Mg. A piece of magnesium wire, a few millims. in length, is for this purpose pushed down into the powdered sample contained in the glass tube; the sodium is carefully freed from rock oil, and rolled out between the fingers to a small cylinder, which is then surrounded by the powdered substance. The best form of carbon is the soot from turpentine, which has been deposited upon the outside of a basin filled with cold water. As soon as the small tube containing the perfectly dry sample has been heated to the point of fusion of the glass, when generally an ignition inside the tube is noticed, it is allowed to cool and then placed upon the porcelain plate covered by a piece of paper and crushed to powder with the knife, for the purpose of further examining the products of reduction.

2. Reduction on Splinters of Charcoal.—In this way the metal can be obtained in small globules, or as a porous mass, from quantities often less than a milligramme of the sample.

A transparent crystal of sodic carbonate is brought near to the outside of the flame, and a common wooden lucifer match then rubbed over two-thirds of its length with the drops of fused salt. If the match is then turned upon its axis through the flame, the carbonized wood becomes surrounded with a crust of solid sodic carbonate, which, on heating in the zone of fusion, melts, and is absorbed by the carbon. A splinter of charcoal is thus obtained, which is prevented from burning by its glaze of soda. A mixture of the substance is then made with the knife upon the hand with one drop of the melted soda-crystal, and a portion of this, of the size of a mustard-seed, placed upon the point of the splinter. As soon as this has been melted in the lower oxidizing flame, it is passed through a part of the dark interior zone to the hotter portion of the lower reducing flame. The point at which the reduction occurs is easily seen by the violent effervescence of the soda; and this is after a time stopped by bringing the splinter into the dark zone. In order to isolate the reduced metal, the end of the splinter is broken off and rubbed up with a few drops of water in a small agate mortar, when the metallic particles are generally visible without removal of the carbon. For further examination, the carbon and soda can be easily removed by several careful washings, and the particles transferred to a small piece of curved glass cut out from an old flask,* in which they are again washed by decantation, the last drops of water removed by suction with a piece of filter-paper, and the metallic particles dried at a moderate heat. A few tenths of a milligramme of the metal is generally sufficient to yield a solution with which all the characteristic precipitations can be accomplished, the reagents being contained in capillary glass threads, dropped into the solution by the milligramme, and the effect thus produced ascertained by examination with a lens. Iron, cobalt, and nickel, which do not fuse to globules on the splinter, are withdrawn from the agate mortar by means of the point of the magnetized blade (Fig. 21, *b*), washed with water, and dried high above the flame on the point of the knife. If the blade be then tightly drawn between the upper part of the thumb and the lower part of the first finger, and if the point of the blade be then approached to the metallic particles on the finger, they jump from the hand to the blade, forming a brush-like bundle, which can be conveniently examined by the lens,

* Watch-glasses crack much too readily to be used for such experiments.

and by touching with a melted borax bead can be transferred in suitable quantities. The portion of metal remaining on the knife is rubbed on to a small piece of filter-paper, a drop of acid added, and the paper warmed over the flame so as to allow the metal to dissolve; this solution can then be further examined with various reagents.

3. Films upon Porcelain.—Those volatile elements which are reduced by carbon and hydrogen can be deposited from their compounds as films on porcelain, either in the elementary state or as oxides. Such films can be extremely easily converted into iodides, sulphides, and other compounds, and thus may be made to serve as most valuable and characteristic tests. The films are composed in the centre of a thick layer, which on all sides gradually becomes thinner until the merest tinge is reached; it is therefore necessary to distinguish between "thick" and "thin" parts of the films. Both exhibit in their variation of thickness all the tints of color characteristic of the substance under different circumstances of division. One-tenth up to one milligramme is in many cases sufficient for these reactions. Many surpass Marsh's arsenic test in delicacy and certainty, and approach in this respect the spectrum-analytical methods.

The following films can be obtained:

(*a*) **Metallic films** are prepared by holding in one hand a particle of the substance on an asbestos thread in the upper reducing flame, which must not be too large, whilst with the other hand a glazed porcelain basin, 1 to 2 decimetres in diameter, filled with cold water, is held close above the asbestos thread in the upper reducing flame. The metals separate out as dead-black or brilliantly-black films of varying thickness. Even Pb., Sn., Cd., and Zn. yield in this way films of reduced metal, which by mere inspection cannot be distinguished from the soot separated out on the porcelain by a smoky flame. By means of a glass rod, these films can be touched with a drop of dilute HNO_3, containing about 20 per cent. of real acid; and the various degrees of solubility of the films serves as a distinguishing characteristic.

(*b*) **Oxide films** are obtained by holding the porcelain basin filled with water in the *upper oxidizing flame*, the rest of the operation being the same as in the production of the metallic films. If only a very small quantity of the sample can be employed, care must be taken to lessen the size of the flame, in order that the volatile products may not be spread over too large a surface of porcelain.

The film of oxide is examined as follows:

(α) The color of the thick and thin film is carefully observed.

(β) The reducing action or otherwise of a drop of stannous chloride is noted.

(γ) If no reduction occurs, NaHO. is added to the stannous chloride until the precipitated hydrate redissolves, and then it is to be observed whether a reduction occurs.

(δ) A drop of perfectly neutral silver-nitrate is rubbed over the film with a glass rod, and a current of ammoniacal air is blown over the surface from a small wash-bottle containing ammonia solution, and having the mouth-tube dipping under the liquid and the exit-tube cut off close below the cork. If a precipitate is formed, the color is observed, and the solubility or alteration, if any, noticed, which occurs when the current of alkaline air is continued, or when a drop of ammonia-liquor is added.

(*b*) **Iodide films** are simply obtained from the oxide-films by breathing on the latter upon the cold basin, which is then placed upon the wide-mouthed, well-stoppered glass (Fig. 26), containing fuming hydriodic acid and phosphorous acid derived from the gradual deliquescence of phosphoric triiodide. When the mixture no longer fumes, owing to absorption of moisture, it is easy to render it again fuming by adding a little phosphoric anhydride. Other films, often containing both iodides of a metal, and therefore frequently less regular in color and appearance, may be easily obtained by smoking the oxide film with a flame of alcohol containing iodine in solution placed upon a bundle of asbestos threads and held under the basin. If any iodine be condensed on the basin with the HI., it can easily be removed by gentle warming and blowing.

Fig. 26.

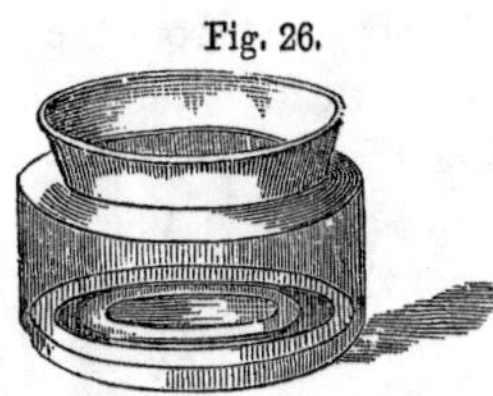

The examination of the film is conducted as follows:

(α) The solubility of the film is examined simply by breathing upon it when the basin is cooled; the color then either changes or entirely disappears, the film being dissolved in the moisture of the breath. If the basin be gently warmed, or if it be blown upon for some distance, the film again becomes visible by the evaporation of the moisture in the current of air.

(β) The ammonia compound of the iodide is formed by blowing ammoniacal air upon it, and noticing whether the color of the thick and thin films alters quickly, slowly, or not at all. The different colors reappear at once if the basin be held for a few moments over an open bottle containing fuming HCl.

(γ) The iodide films generally give the same reactions

as the oxide films with silver-nitrate and ammonia, with stannous chloride, and with caustic soda.

(*b*) **The Sulphide film** is most easily obtained from the iodide film by blowing upon it a current of air saturated with ammoniac sulphide, and removing the excess of sulphide by gently warming the porcelain. It is advisable to breathe on the film from time to time whilst the current of sulphuretted air is being blown on the basin. The experiments to be made with this film are:

(α) The solubility or otherwise in water is ascertained by breathing on it, or by addition of a drop of water. The sulphides often possess the same color as the iodide films; they may, however, generally be distinguished by their insolubility on breathing.

(β) The solubility of the sulphide in ammoniac sulphide is ascertained by blowing or dropping.

(*b*) **Films on Test-tubes.** — Under certain circumstances it is advisable not to collect the film on porcelain, but upon the outside of a large test-tube (Fig. 22, D); this method is especially used when it is needed to collect larger quantities of the reduction film for the purposes of further examination. The fine asbestos thread with the sample of substance is held on the glass tube (*b*) before the lamp so that it is placed at the height of the middle of the upper reducing flame, and the test-tube fixed so that the lowest point is just above the end of the asbestos thread. If the lamp be now pushed under the test-tube, the substance and the asbestos thread are in the reducing flame. By repeating this operation the film can be obtained of any wished-for thickness; some pieces of marble are in this case placed in the test-tube, to prevent the water from being thrown out of the tube by percussive boiling.

TABLE OF VOLATILE ELEMENTS WHICH CAN BE REDUCED AS FILMS.

	METALLIC FILM.	OXIDE FILM.	OXIDE FILM WITH STANNOUS CHLORIDE.	OXIDE FILM WITH STANNOUS CHLORIDE AND SODA.	OXIDE FILM WITH SILVER-NITRATE AND AMMONIA.
Te	Black; thin part brown.	White.	Black.	Black.	Yellowish-white.
Se	Cherry-red; thin part brick-red.	White.	Brick-red.	Black.	White.
Sb	Black; thin part brown.	White.	White.	White.	Black; insoluble in ammonia.
As	Black; thin part brown.	White.	White.	White.	Lemon-yellow or reddish-brown; soluble in ammonia.
Bi	Black; thin part brown.	Yellowish-white.	White.	Black.	White.
Hg	Gray non-coherent thin film.				
Tl	Black; thin part brown.	White.	White.	White.	White.
Pb	Black; thin part brown.	Yellow-ochre color.	White.	White.	White.
Cd	Black; thin part brown.	Blackish-brown; thin part white.	White.	White.	White; in the thin parts turns bluish-black.
Zn	Black; thin part brown.	White.	White.	White.	White.
Sn	Black; thin part brown.	Yellowish-white.	White.	White.	White.

TABLE OF VOLATILE ELEMENTS WHICH CAN BE REDUCED AS FILMS — Continued.

IODIDE FILM.	IODIDE FILM WITH AMMONIA.	SULPHIDE FILM.	SULPHIDE FILM WITH AMMONIUM SULPHIDE.	
Brown; disappears for a time on breathing.	Disappears altogether on blowing.	Black to blackish-brown.	Disappears for a time.	Elements whose reduction films are scarcely dissolved in dilute nitric acid.
Brown; does not wholly disappear on breathing.	Does not disappear on blowing.	Yellow to orange.	Orange, and then disappears for a time.	
Orange-red to yellow; disappears on breathing.	Disappears altogether on blowing.	Orange.	Disappears for a time.	
Orange-yellow; disappears for a time on breathing.	Disappears altogether on blowing.	Lemon-colored.	Does not disappear.	
Bluish-brown; thin parts pink; disappear for a time on breathing.	Pink to orange; chestnut colored when blowing.	Burnt-umber-color to coffee-colored.	Does not disappear.	Elements whose reduction films are with difficulty dissolved in dilute nitric acid.
Carmine-colored and lemon-yellow; does not disappear on breathing.	Disappears for a time on blowing.	Black.	Does not disappear.	
Lemon-yellow; does not disappear on breathing.	Does not disappear on blowing.	Black; thin parts bluish-gray.	Does not disappear.	
Orange-yellow to lemon-color; does not disappear on breathing.	Disappears for a time on blowing.	Brownish-red to black.	Does not disappear.	Elements whose reduction films are instantly dissolved in dilute nitric acid.
White.	White.	Lemon-colored.	Does not disappear.	
White.	White.	White.	Does not disappear.	
Yellowish-white.	Yellowish-white.	White.	Does not disappear.	

C. The Reactions of the Elements.

The elements, which can easily be recognized by their flame reactions, are arranged in the following groups and sub-groups according to their behavior in the reducing and oxidizing flames:

A. Elements which are reducible to metal and are deposited in films.

1. Films scarcely soluble in cold dilute nitric acid — Tellurium, selenium, antimony, arsenic.
2. Films slowly and difficultly soluble in cold dilute nitric acid — Bismuth, mercury, thallium.
3. Films instantly soluble in cold dilute nitric acid — Cadmium, zinc, indium.

B. Elements reducible to the metallic state, giving no film:

1. Not fusible to a metallic bead.
   a. Magnetic — Iron, nickel, cobalt.
   b. Non-magnetic — Palladium, platinum, rhodium, iridium.
2. Fusible to metallic beads — Copper, tin, silver, gold.

C. Elements most easily separated and recognized as compounds — Tungsten, titanium, tantalum and niobium, silicon, chromium, vanadium, manganese, uranium, sulphur, phosphorus.

SPECIAL REACTIONS.

1. Tellurium Compounds. — *Flame coloration,* in upper reducing flame, pale-blue, whilst the oxidizing flame above appears green.

Volatilization unaccompanied by any odor.

Reduction film, black, with dark-brown coating, dull or

brilliant; heated with concentrated sulphuric acid, gives a carmine-red solution.

Oxide film, white, scarcely, or not at all, visible; protochloride of tin colors it black, by reason of separated tellurium; nitrate of silver, after ammonia has been blown upon it, yellowish-white.

Iodide film, dark-brown, with brown coating; disappears momentarily when breathed upon, but not when slightly warmed; reappears on exposure to HCl.; blackened by SnCl.

Sulphide film, dark-brown to black; does not disappear when breathed upon; dissolves in NH_4S. blown upon it, and reappears upon warming, or if blown upon with air.

With soda on charcoal splinter gives a telluride of sodium, which, when moistened upon a silver coin, produces a black spot, and if the specimen contains much tellurium, with HCl., diffuses an odor of telluretted hydrogen with the separation of black tellurium.

2. Selenium Compounds. — *Flame coloration*, pure azure-blue.

Volatilizes, burning with the odor of selenium.

Reduction film, brick-red to cherry-red; at one time dull, at another brilliant; gives, when heated with concentrated SO_3, a dirty-green solution.

Oxide film, white; brick-red from separated selenium when SnCl. is dropped upon it; the old color darkened by NaO, HO; with AgO, NO_5 the oxide film gives a white, scarcely visible coloration, which disappears when ammonia is blown upon it.

Iodide film, brown, contains some reduced selenium, and therefore cannot be made to disappear completely, either by breathing upon it or by blowing ammonia upon it.

Sulphide film, yellow to orange-red, insoluble in water, soluble in NH_4S.

With soda on charcoal splinter gives selenide of sodium, which produces, with a drop of water, a black spot upon a silver coin, and moistened with HCl., if the quantity is not too small, gives the odor of selenuretted hydrogen with separation of red selenium.

3. Antimony Compounds.—*Flame coloration*, by treatment in the upper reducing flame, pale-green, unaccompanied by any smell.

Reduction film, black; sometimes dead, sometimes bright.

Oxide film, white; moistened with a perfectly neutral solution of AgO,NO_5, and then blown on by ammoniacal air, it gives a black spot which does not disappear in NH_4O. If the film be first placed over bromine vapor the reaction cannot be obtained, owing to the oxidation of SbO_3 into SbO_5. It is unaltered by SnCl., either with or without NaO, HO.

Iodide film, orange-red, disappearing by breathing, and reappearing by blowing or warming; blown on with ammoniacal air it disappears, but does not return. Generally it gives the same reactions as the oxide.

Sulphide film, orange-red. The film is difficult to blow away with NH_4S; returns on blowing with air; insoluble in water.

With soda on charcoal splinter gives no black stain on silver, but yields a white, brittle, metallic bead.

4. Arsenic Compounds.—*Flame coloration* in upper reducing flame pale-blue, giving the well-known arsenical smell.

Reduction film black, dead, or brilliant; thin film brown.

Oxide film, white; touched with a perfectly neutral solution of $AgO.NO_5$, and then blown with ammoniacal air, it gives a canary-yellow precipitate, soluble in NH_4O. Together with this yellow precipitate, a brick-red one of silver-arseniate occurs when the film has previously been treated with bromine vapor. SnCl., with and without soda, produces no change.

Iodide film is deep-yellow, disappears on breathing, but returns on drying; disappears in ammoniacal air, and does not return; reappears unaltered after the action of HCl.

Sulphide film, lemon-yellow, disappears easily on blowing with NH_4S., and reappears on warming or blowing; insoluble in HO., and does not disappear by blowing upon it.

Reduction on charcoal splinter yields no metallic bead.

5. Bismuth Compounds.—*Reduction film,* black, dead, or brilliant; thin portion of film, brownish-black.

Oxide film, light-yellow, unaltered by $AgO.NO_5$, with or without ammonia; gives no reaction with SnCl., but yields black precipitate of BiO_2 on addition of NaO.HO.

Iodide film is very characteristic, and remarkable for the number of tints which it assumes. The thick part is of a brown or blackish-brown color, with a shade of lavender-blue; the thin film varies from flesh-color to light-pink; it easily disappears on breathing, and appears again on blowing. In a stream of ammoniac air it passes from pink to orange, and on blowing or warming it again attains a chestnut-brown color; it resembles the oxide film in its behavior with SnCl. and NaO.HO.

Sulphide film is of a burnt umber color; the thin parts are of a lighter coffee-brown color; does not disappear on blowing, and is not soluble in NH_4S.

On charcoal splinter with soda the bismuth compounds are reduced to a metallic bead, yielding, when rubbed in the mortar, bright, shining, yellowish splinters of metal soluble in NO_5. The solution gives, with SnCl. and NaO. HO., black BiO_2.

6. Mercury Compounds. — *Metallic film* is mouse-gray, noncoherent, and spreads over the whole basin. To obtain small traces of Hg. in the reduced state, the sample is mixed with soda and KO,NO_5 and filled into a thin test-tube five to six millims. wide and ten to twenty millims. long. This is held by a Pt. wire in the flame, whilst the bottom of the basin, filled with cold water, is placed close above the open end of the tube. If the quantity of Hg. is considerable, it collects in the form of globules, which can be seen with a lens, and which can be collected into larger drops by wiping the basin with a piece of moistened filter-paper.

Iodide film is obtained by breathing on the metallic film, and then placing it over the vessel (Fig. 26 of the Plate) containing moist Br. It first becomes black, and then disappears, but not until after some time; HgBr. is formed. If the basin be now placed above the vessel of fuming HI., a very characteristic carmine-colored film of Hg_2I. is produced; this is often accompanied by HgI., but neither of these disappear on breathing, nor when blown with ammoniacal air.

Sulphide film, black; not altered by breathing or by blowing with NH_4S.

7. Thallium Compounds. — Since the minutest trace of this element can be recognized by means of the spectroscope, it will seldom be detected in any other way.

Flame coloration, bright grass-green.

Metallic film, black, with coffee-brown coating.

Oxide film, colorless; unchanged by SnCl., or NaO. HO., also with $AgO.NO_5$ with or without NH_4O.

Iodide film, lemon-yellow; insoluble in NH_4O.

Sulphide film, obtained from the oxide film, black, with livid coating; insoluble in NH_4S.

On charcoal with soda, reducible to a white ductile grain.

8. Lead Compounds. — *Flame coloration,* pale-blue.

Reduction film, black, dead, or brilliant.

Oxide film, bright yellow-ochre colored; stannous chloride gives no reaction even on addition of NaO.HO; $AgO.NO_5$ does not produce any reaction, either alone or on addition of NH_4O.

Iodide film, orange- to lemon-yellow, insoluble on breathing or on moistening; disappears on blowing with ammoniacal air, and again appears on warming.

Sulphide film, brownish-red to black; by blowing or moistening with NH_4S. it remains unaltered.

On charcoal splinter with soda gives a gray, very soft, ductile metallic bead, which is slowly but completely soluble in $HO.NO_5$, yielding a white, easily crystallizable salt, soluble in HO., and precipitated as a white powder on addition of SO_3 from a capillary tube.

9. Cadmium Compounds. — *Metallic film,* black; the thin parts, brown.

Oxide film, brownish-black, shading off through brown to a white invisible film of suboxide, which is not changed by SnCl., alone or with soda; $AgO.NO_5$ produces a blackish-blue coloration of reduced metal, which is very characteristic and does not disappear on addition of NH_4O.

Iodide film, white; no change produced by NH_4O.

Sulphide film, lemon-yellow, insoluble in NH_4O.

Reduction on charcoal splinter with soda. — The metal, owing to its volatility, can be obtained, only with difficulty, as a silver-white ductile bead.

10. Zinc Compounds. — *Reduction film*, black; in the thin parts, brown.

Oxide film, white, and therefore invisible. To test it, a square centimetre of filter-paper moistened with NO_5 is rubbed over the surface, and then rolled up on two rings on fine platinum wire, three millimètres in diameter, and burnt. If the paper is burnt in the upper oxidizing flame at as low a temperature as possible, the ash forms a small solid mass about a square millimètre in area, which can be ignited without fusion, and becomes yellow on gently heating, and appearing white on cooling. If this be moistened with a few milligrammes of very dilute CoO. NO_5 solution and ignited, it appears of a beautiful green color on cooling; the same reaction can be effected with the metallic film.

Iodide film, white; not clearly recognizable either alone or after ammonia has been blown upon it.

Sulphide film, also white, and not easily recognized either alone or when moistened with NH_4S.

Reduction on charcoal splinter does not proceed on account of volatility of the zinc.

11. Indium Compounds. — Detected with most ease and certainty with the spectroscope.

Flame coloration, intense; pure indigo-blue.

Metallic film, black, with brown coating; at one time dull, at another brilliant. Disappears instantly with NO_5.

Oxide film, yellowish-white; scarcely visible; give no reactions with SnCl. and AgO.NO_5 solution.

Iodide film, also yellowish; nearly white, invisible, if weak, with and without ammonia.

Sulphide film, also yellowish; nearly white, scarcely visible. Unchanged by NH_4S.

Reduction on charcoal splinter with soda takes place with difficulty, and affords silver-white, ductile globules, slowly soluble in HCl.

12. Iron Compounds.—*Reduction on charcoal splinter* gives no metallic bead or ductile lustrous particles; the finely divided metal forms a black brush on the end of the magnetized knife-blade; this, when rubbed off on paper and dissolved in a drop of aqua regia, yields a yellow spot when warmed over the flame, which, when moistened with ferrocyanide of potassium, gives a deep coloration of *Prussian Blue.* The yellow spot moistened with NaO, HO., and then held for a few moments in a vessel with bromine vapor, gives, on a second addition of soda, no coloration of a higher oxide.

Borax Bead. — In the oxidizing flame, when hot, yellow to brownish-red; when cold, yellow to brownish-yellow; reducing flame, bottle-green.

13. Nickel Compounds.—*Reduction on the charcoal splinter.* — On pulverizing the charcoal, white, lustrous, ductile, metallic particles are obtained, forming a brush on the magnetized blade. The metal, dissolved in NO_5 on paper, gives a green solution, which on moistening with soda, exposure to bromine vapor, and second addition of soda, give a brownish-black spot of Ni_2O_3. The ash of the paper, from which the soda has been washed out, can be used for the borax bead test.

Borax bead. — Oxidizing flame, grayish-brown, or dirty-violet. Upper reducing flame, gray, from reduced Ni., which often collects to a spongy mass of metal, rendering the bead colorless.

14. Cobalt Compounds.—*Reduction on charcoal splinter.*

— By pulverizing the charcoal, white, ductile, lustrous, metallic particles are obtained, which form a brush on the magnetic blade. The metal, rubbed off on to paper, gives a red solution when moistened with NO_5; this yields a green color on addition of HCl. and drying, which disappears again on moistening. The paper moistened with soda, brought into bromine-vapor and again moistened with soda, yields a brownish-black spot of Co_2O_3. This reaction is plainly seen with a few tenths of a milligramme of metal. The paper can also be used, after washing out the soda and burning, for the coloration of the borax bead.

Borax bead. — Deep-blue bead in the oxidizing flame, which does not change in the lower reducing flame. When treated for a considerable time alone, or better still with platin-chloride of ammonium, in the most energetic upper reducing flame, is completely decolorized, but only after long treatment, with the separation of cobalt or platin-cobalt.

15. Palladium Compounds. — These are reduced upon fine platinum wire, with soda, in the upper oxidizing flame, to a gray mass, similar to platinum sponge, which, when rubbed in an agate mortar, gives shining, ductile, metallic scales. The scales, rinsed and dried upon a piece of glass plate, dissolve in NO_5, with reddish-brown color.

If a small drop of a solution of cyanide of mercury is added to the liquid, a white flocculent precipitate is obtained, which dissolves in NH_4O. when dropped upon it. After evaporation and boiling with aqua regia, the liquid, evaporated to a small drop, gives a dirty orange-yellow, crystalline precipitate of ammonio-chloride of palladium.

Solution of palladium is colored blue, green, and brown, by SnCl., according to the amount used.

16. Platinum Compounds. — These give when ignited upon platinum wire with soda in the upper oxidizing flame also a gray spongy mass, which, by rubbing in an agate mortar, is converted into shining, silver-white, ductile metallic scales. These are insoluble either in NO_5 or HCl. alone, but with aqua regia give a bright-yellow solution, if the platinum is pure; if it contains rhodium, iridium or palladium, they give a brownish-yellow solution. When solution of cyanide of mercury is added to the solution, and ammonia blown upon it, no flocculent white precipitate is formed, but immediately a bright-yellow crystalline precipitate of platin-chloride of ammonium. SnCl. colors solutions of platinum yellowish-brown.

17. Iridium Compounds. —These, ignited in the upper oxidizing flame with soda, are likewise reduced to metal, which, when rubbed in an agate mortar, forms a gray powder, without lustre, and not in the least degree ductile. This is insoluble in nitric acid, hydrochloric acid, and aqua regia.

18. Rhodium Compounds.—These are only distinguished from the iridium compounds by the fact that the metallic powder, insoluble in aqua regia, when fused with bisulphate of potassa, is partially oxidized, and affords a rose-red solution.

19. Osmium Compounds. — These give in the oxidizing flame volatile osmic acid, of a pungent odor, similar to chlorine, and which irritates the eyes.

20. Gold Compounds. — If only traces of gold are present, mixed with a considerable quantity of gangue, it can only be concentrated and detected according to the old processes for detecting gold. Otherwise even a few tenths of a milligramme can be recognized by reduction with soda on a charcoal splinter. The yellow, shining, ductile, metallic grain, obtained in this way, can be reduced to spangles having the lustre of gold, by rubbing in an agate mortar. These are insoluble in HCl. or NO_5, but give, rather readily, with aqua regia, a bright-yellow solution. If this is soaked up into a piece of blotting paper, and touched with SnCl., purple of Cassius is formed. What remains upon the glass is colored brown by a solution of $FeO.SO_3$, by reason of separated gold, whilst the liquid appears blue by transmitted light.

21. Silver Compounds. — If silver occurs only in traces in slags or complex ores, it can only be detected by the well-known method of cupellation. If, however, the silver compound is not mixed with a very large amount of foreign matter, it can be detected in very minute quantities by reduction with soda on the charcoal splinter. The white ductile beads dissolve easily on warming in dilute NO_5, and yield AgCl. with HCl., which can then readily be recognized by its behavior with NO_5 and NH_4O. Less than one-tenth of a milligramme of silver can thus be easily detected with certainty.

22. Copper Compounds. — *On the charcoal splinter with soda* the copper compounds yield a ductile lustrous metallic bead, easily recognizable by its red copper color. By rubbing in the mortar, flat metallic particles are obtained, which can be readily washed, and are easily soluble in NO_5. The blue solution, absorbed on filter paper, yields a brown stain on addition of ferrocyanide of potassium. Instead of acting upon a metal in a curved glass, it may be dissolved by moistening paper upon which it is placed with NO_5.

With borax on platinum wire. — Blue bead, not altered to cuprous oxide when heated in the lower reducing flame alone, but on addition of very little tin-salt forms a reddish-brown bead. If this bead be frequently oxidized and reduced in the flame, a ruby-red transparent bead is obtained; this occurs most readily when the bead is allowed to oxidize very slowly.

23. Tin Compounds. — *On the charcoal splinter* the Sn. compounds are easily reduced to white, lustrous, ductile, metallic beads. The flattened particles, transferred to the curved glass, slowly dissolve in HCl.; and the solution, when absorbed by paper, gives a red precipitate with selenious, and a black precipitate with tellurous acid dissolved in HCl. If to the solution a trace of bismuth-nitrate be added, an excess of soda gives a black precipitate of BiO. The metal, acted on by NO_5, yields a white powder of insoluble SnO_2.

A borax bead, containing enough CuO. to render it faintly blue, serves as a delicate test to ascertain with certainty the presence of a trace of a Sn. compound, as the bead, placed in the lower reducing flame, turns reddish-brown, or forms a clear ruby-red glass.

24. Molybdenum Compounds. — *On a charcoal splinter with soda* are reduced, with great difficulty, to a gray powder. In the same way some compounds give in the upper reducing flame a film on porcelain, which it is difficult to obtain. Molybdenum is best recognized as follows:

a. The sample is finely pulverized with the knife on the porcelain plate, is mixed on the hand with soda, obtains a pasty state by fusion. The mixture is then transferred to a spiral of fine platinum wire and fused in the flame. The liquid fused mass is then knocked off the wire and allowed to fall upon the plate, when it is digested with two or three drops of water, and the clear liquid above the sediment is soaked up into three or four strips of filter-paper, not too fine, several millimètres broad. One of these strips, moistened with HCl., does not change color, but with a drop of ferrocyanide of potassium is changed to reddish-brown.

If one of these strips is gradually moistened with a few milligrammes of SnCl., it either becomes blue in the cold or upon warming; if it becomes yellow or yellowish-brown, more of the solution of the test specimen must be added by means of a capillary pipette, in order to cause the blue color to appear.

A drop of NH_4S., placed upon the third strip, produces a brown color, and on addition of HCl., a brown precipitate, whereupon the paper often becomes blue on the edge of the precipitate.

The yellow phosphate precipitate produced by the nitric acid solution of molybdate of ammonia can also be readily obtained. The slightly borax bead is colorless in the oxidizing flame; when it contains more molybdenum bluish, enamel-like; in reducing flame dark, by reason of reduced molybdenum.

25. Tungsten Compounds. — The reduction of tungsten can also be made on charcoal splinter with soda, but is not adapted to the separation or detection of the metal. The compounds are, therefore, treated in the manner just given for molybdenum, by soaking up the fluid, after fluxing with soda, with strips of filter-paper.

One strip moistened with HCl. remains white, but if heated turns yellow; moistened with ferrocyanide, unchanged. A second strip touched with SnCl. is colored blue even when cold or on warming. A drop of NH_4S. causes no precipitate alone or with HCl.; the paper, however, becomes blue or greenish on warming.

26. Titanium Compounds. — These give a colorless bead with microcosmic salt in the oxidizing flame, which turns a pale amethystine color in the reducing flame. On addition of $FeO.SO_3$ the bead assumes in the reducing flame the peculiar red color of venous blood, whilst in the oxidizing flame the light brown color of Fe_2O_3 can be obtained at pleasure. The titanium compounds form with soda a bead, which at first effervesces, and when hot is colorless and transparent, but on cooling becomes opaque. If to the hot bead SnCl. be added, and if it then be heated in the lower reducing flame, a gray mass is formed, which dissolves on heating in HCl., yielding a pale amethystine-colored solution.

27, 28. Tantalum and Niobium Compounds.—These manifest the same reactions as titanium.

29. Silica Compounds.—The silicates, when treated in the oxidizing flame with carbonate of soda, dissolve, more or less, with effervescence. The hot fused, when moistened with SnCl. and thoroughly ignited, does not give a trace of a blue coloration, when evaporated upon the porcelain plate, whereby the silica may be distinguished from titanic, tantalic, and niobic acids. It likewise fails to give the blood-red coloration in a bead with sesquioxide of iron, produced by these acids. The fluxed mass, if water and acetic acid are carefully added, and then evaporated on the porcelain plate, separates gelatinous hydrate of silica. Fine splinters of silicate give, upon fusing in the bead of microcosmic salt, a gelatinous skeleton of silica, floating in the fused or cooled bead.

30. Chromium Compounds.—In platinum spiral with soda the compounds when fluxed, with the repeated addition of nitrate of potassa, give a bright-yellow mass, which, when knocked off on to the porcelain plate and crushed, give a bright-yellow solution. If this solution is decanted from the residue, and acetic acid added, it becomes yellowish-red, and gives with lead salts, when it is soaked up by strips of filter paper, a yellow precipitate; with solutions of salts of oxide of mercury, a red one; and with AgO,NO_5, a reddish-brown one.

With NH_4S., also by evaporation with aqua regia upon the porcelain plate, the solution becomes green; likewise with SnCl. The *borax bead* becomes emerald-green in the oxidizing flame, and does not change this color in the reducing flame.

31. Vanadium Compounds.—Treated with soda and nitrate of potassa in a platinum spiral yield a bright-yellow mass, the solution of which, on addition of AgO,NO_5 and acetic acid, yields a yellow precipitate. The fused mass, when evaporated with aqua regia, gives a yellow instead of a green solution, which becomes blue on addition of SnCl. If much vanadium is present, the solution gives a yellowish-brown solution or precipitate on addition of concentrated cold HCl. In the *borax bead* these compounds give a yellowish-green color in the oxidizing flame; in the reducing flame, a green color.

32. Manganese Compounds.—*Borax bead.*—Amethyst in the oxidizing flame; colorless in the reducing flame.

With carbonate of soda on platinum wire, a bead is formed green after cooling, especially easily after addition of nitrate of potassa. Water extracts a green solution from it, which becomes red after the addition of acetic acid, and then, often with the separation of brown flakes, becomes colorless.

33. Uranium Compounds give a yellow bead in the oxidizing flame, which becomes green in the reducing flame, especially on addition of SnCl. These colors closely resemble those of the iron compounds, but may easily be distinguished, at least if no other coloring metallic oxide is present, by the fact that the uranium bead, when incandescent, emits a bluish-green light, analogous to that which the uranium compounds exhibit when fluorescing. Beads of lead-oxide, stannic acid, and a few other substances, exhibit a similar phenomenon when incandescent, but they do not yield, like uranium compounds, a colored bead on cooling.

Heated gently on the platinum spiral with $KO,2SO_3$, the insoluble uranium compounds can be decomposed. The melted mass is powdered with a few particles of crystallized carbonate of soda, and the moistened mass is absorbed by filtering paper. A brown spot is formed by the addition of a drop of ferrocyanide of potassium to the moistened paper.

34. Phosphorus Compounds.—These may easily be detected in its compounds, even when they are mixed with large quantities of other substances, as follows:

The sample, having been ignited, is rubbed fine on the porcelain plate (see Fig. 25), and is then introduced into a small glass tube of the thickness of a straw; into this tube, which is closed at the bottom, a piece of magnesium wire, about one-fourth of an inch in length, is placed so that it is covered by the powder. On heating the tube, magnesic phosphide is formed with incandescence. The black contents of the tube powdered on the plate give, on moistening with HO., the highly characteristic smell of H_3P. A piece of Na. can be equally well used if the Mg. wire cannot be procured.

If it has been ascertained that the sample does not yield any film on porcelain in the upper oxidizing flame, the phosphates may be recognized by heating on platinum with borax and a thin piece of iron wire in the hottest part of the reducing flame, when a bright molten bead of iron phosphide is obtained, which can be extracted with the magnetized knife on crushing the bead under paper.

35. Sulphur Compounds. — *On a charcoal splinter with soda,* in the lower reducing flame, they give a fused mass, which, moistened upon a piece of silver, blackens it. Since selenium and tellurium produce the same reaction, the absence of these substances must be determined by the absence of a tellurium or selenium spot upon porcelain.

When only metallic sulphides are to be considered, and not sulphates, it will answer simply to heat the test specimen in the flame to detect sulphur by the odor.

It will be sufficient to show the advantages of the methods described by one or two examples.

a. A mixture of sulphide of arsenic, sulphide of antimony, and sulphide of tin. — If in a mixture of these three sulphides, containing only traces of Sb. and Sn., they are separated according to the ordinary rules of qualitative analysis, by dissolving in alkaline sulphides and reprecipitation with acids, the detection of these two metals by the regular tests is extremely uncertain and troublesome. According to the following method the detection of these metals is rendered easy and certain when the proportion of Sn. is only a few thousandths, and that of the Sb. only a few hundredths, of the total weight of the mixture.

Three decigrammes of the sulphides are roasted on a curved piece of glass * small enough to be altogether surrounded by the flame, and the residue, weighing only a few milligrammes, is scraped together with the knife. The moistened mass is then collected on the end of a thread of asbestos and a strong metallic film obtained on the test tube. In order to prevent the deposition of any carbon with the metals, which would act injuriously in the

* Pieces of a thin chemical flask are also best for use in this case.

subsequent operations, the upper reducing flame is made so small that the luminous point is only just visible. The film is next dissolved in a drop or two of NO_5 in the curved rim (Fig. 25, of the Plate), and the solution evaporated below its boiling-point by gently warming and blowing, so as to obtain the solid residue in as small a space as possible. A drop of neutral silver solution is now brought on to the residue at the moment when it becomes solid; and on blowing with ammoniacal air a characteristic black stain is formed, whilst the reaction of As. is also generally noticed.

In order to detect Sn., a few scarcely visible particles of the roasted sulphides are fused on to a borax bead which has been very slightly tinted with cupric oxide. If the bead is now brought into the lower reducing flame, it becomes a ruby-red color from reduced cuprous oxide. If the oxide be present in too large a quantity, the bead can be obtained transparent by the process described under the reactions of the copper compounds. This reaction can only be obtained in the lower reducing flame of the non-luminous gas-lamp, as in the ordinary blow-pipe flame the cupric oxide is reduced to cuprous oxide without the presence of tin-salt.

b. Black tellurium, containing tellurium, selenium, antimony, gold, lead, and sulphur.—After the sulphur has been detected by the smell by roasting, the metallic film is obtained on a test-tube, which is then placed inside a wider and shorter tube containing a few drops of *concentrated* sulphuric acid, so that the metallic film is surrounded by the acid. If the temperature be now gradually raised, the presence of tellurium is at once ascertained by the formation of a bright carmine color. If the temperature be still further raised, the tellurium oxidizes, and the olive-green color of selenium becomes visible; the cooled solution, on dilution with water, then no longer exhibits the black precipitate of tellurium, but is colored yellowish-red with the selenium. If this is present in small traces only, it can be best detected by looking down the length of the test-tube upon a sheet of white paper. As common commercial SO_3 often contains traces of selenium, it is well to make a trial experiment first. The antimony is detected as in preceding example. To detect the lead and gold, a sample is reduced on the charcoal splinter, the beads of the alloy are washed into a curved glass, and the flattened and dried metallic particles treated with rather strong NO_5 as long as anything dissolves. The acid is then evaporated off, and the soluble portion of the residue dissolved in a drop or two of water. The solution is brought on to a second curved glass by means of a capillary pipette, and the characteristic precipitate of sulphate of lead obtained with SO_3. The gold left undissolved as a brown powder, is completely washed by frequent addition of water and removal of the same with the capillary pipette. A portion of the dried particles of gold fused on a charcoal splinter with soda, yields in the mortar bright-yellow, golden particles, which may be dissolved in aqua regia and tested with SnCl. A centigramme of the sample is sufficient in experienced hands for all these reactions.*

SPECTRUM ANALYSIS.

The second method—that of *spectrum analysis*, discovered by Kirchhoff and Bunsen, consists in letting the rays of the colored flame, after passing through a narrow slit, traverse a prism, and in observing the spectrum thus produced by means of a telescope. Each of the metals which

* For Tables of Volatile Elements which can be reduced as films, see pp. 210, 211.

give color to the flame, thus yields a peculiar spectrum, formed in some cases, as in that of baryta, of many contiguous colored lines; in others of two more distant lines of different color, as shown by lithia; or, again, of a single line, as in case of soda and thallium. These spectra are characteristic in two respects, viz.: 1, in the definite color of the spectrum lines; and 2, in the invariable relative position they occupy.

The last-named fact enables us to detect, without difficulty in most cases, all the spectrum-giving ingredients of a mixture. Thus, when potassa, soda and lithia-salts are brought together into the *spectroscope*, the lines characteristic of each metal appear in the utmost purity at one view. Very minute traces of some elements do not, however, exhibit their spectra in presence of large quantities of other substances.

The methods of obtaining the spectra of the elements, or their compounds, vary according to their volatility. The instrument or spectroscope used varies according to the degree of accuracy which the observations require. Thus, for the detection of some of the more commonly occurring substances, a rough apparatus with one prism will suffice, whereas for exact experiments, a much more powerful and perfect instrument is needed.

Every spectrum apparatus or spectroscope, exclusive of the source of light, is composed of an adjustable slit, a contrivance (collimating lens) for rendering the rays parallel that have passed through the slit, and a prism. All light except that under examination must be excluded from the prism, and therefore the slit, prism, and lenses are enclosed in a tube, or, if the prism be too large, the latter is fitted with a separate cover. As the spectrum on emerging from the prism is but little longer than the width of the slit, and only becomes of some length as the distance from the prism increases, a magnifying glass is introduced, in order that the eye, though at but a small distance from the prism, may see the spectrum of a sufficiently large size, and the spectrum, therefore, is not observed with the naked eye, but through the medium of a telescope of moderate power. This telescope is also necessary for enabling the eye to receive the whole of the light passing from the collimating lens through the prisms.

In order to know the exact position of the lines another tube is added, in the end of which is placed a scale photographed on glass, and which is so reflected that the observer sees the scale and spectrum at the same time.

A small glass prism may also be attached to the end of the tube through which the colored flame is admitted, so that the spectra of two flames may be examined and compared at the same time.

Not only the number of the spectrum lines of a substance, but also the degree of their intensity, is deserving of careful attention. As the brilliancy of the lines increases with the temperature, so, as a rule, it is those lines which are particularly prominent at a high degree of temperature that are the first to appear at a low temperature. These are best suited for the recognition of a substance, and are therefore called *characteristic* lines. Such lines, according to their brightness, are designated in each substance by the letters of the Greek alphabet, α, β, γ, δ, etc., being affixed to the symbol of the element.

The table at the beginning of the book exhibits the spectra of some of the more commonly occurring and easily recognized elements.

Potassium. — All the volatile compounds of potassium, when placed in the flame, give a widely extended, continuous spectrum, which consists of two chief lines; one line $K\alpha$, situated in the outermost red, and a second line, $K\beta$, situated far in the violet rays towards the other end of the spectrum.

When vapor of potassium is heated in the electric spark, several other lines make their appearance. Before testing potash silicates they should be ignited with carbonate of soda, as it does not interfere with the reaction. Orthoclase, sanidin, and adularia may be easily distinguished from albite, orthoclase, anorthite, and labradorite. If only a trace of potash is present, the silicate should be heated with fluoride of ammonium in a platinum dish, and the residue placed in a flame with a platinum wire.

Sodium. — The yellow line, $Na\alpha$, is the only one which appears in the sodium spectrum as seen in the flame with the ordinary spectroscope. With powerful instruments it is seen to be double. The line α is remarkable for its definite form and brightness, and is produced by all the natural compounds of sodium.

Lithium. — The salts of this metal give a bright line in the red, $Li\alpha$, and another much less distinct in the orange, $Li\beta$. With high heat and strong prisms a blue line also appears. All lithium compounds give the reaction, and often only require to be held in the flame, as lepidolite, petalite, etc. If the amount is very small in the silicate it should be digested and evaporated with fluoride of ammonium, a little sulphuric acid added, again evaporated, and the residue treated with alcohol. The solution in alcohol is evaporated to dryness, the mass again treated with alcohol, and the liquid dried in a glass capsule. The crust formed can be placed in the flame with a platinum wire.

Strontium. — The spectra of the alkaline earths are more complex than those of the alkalies. Strontium gives eight very distinct lines — six red, one orange, one blue. The orange line, Srα, close to the sodium line, the two red lines, Srβ and γ, and the blue line, Srδ, are the most important. The chloride gives the best reaction, while the non-volatile compounds give none. The sulphate must be reduced to sulphide, by holding the bead in the reducing flame, and the silicates must be fused with carbonate of soda, powdered and washed with water by decantation. The insoluble carbonate thus obtained is moistened with hydrochloric acid, and will then give a distinct reaction. The strontium lines do not interfere with the indications of the alkalies.

Calcium. — The spectrum of this metal is easily distinguished from all the foregoing by the green line, Caβ, and by the orange, Caα. A feeble line is also seen in the violet with a powerful instrument, and other lines with increased heat. The chloride gives the best reaction. The non-volatile compounds must be decomposed by hydrochloric acid, and the silicates by fluoride of ammonium. The composition of calcareous rocks and minerals is easily found.

Barium. — The complicated spectrum of this metal is distinguished by its green bands, of which Baα and β are the most important. The haloid salts and more common natural compounds are recognized by holding them in the flame. The silicates must first be treated with hydrochloric acid or fused with carbonate of soda and then dissolved in acid. If barium and strontium occur in small quantities with calcium, the carbonates obtained by fusion are dissolved in nitric acid and the dried salt exhausted with alcohol. The residue contains only barium and strontium, which can generally be detected.

Rubidium.—The continuous spectrum is not so extended as that of potassium. The α and β are most brilliant and best suited for the recognition of the metal. The lines δ and γ are less intense but still very characteristic. These lines and even others appear with not only the volatile chloride, nitrate, etc., but even with the silicates.

Cæsium.—The spectrum is characterized by two lines, Csα and Csβ, both brilliant and well-defined. The absence of any line in the red distinguishes this from the two previous spectra. With an intense light, yellow and green lines may be seen in the continuous parts of the spectrum. Rubidium and cæsium can be easily separated from lithium and sodium by means of bichloride of platinum, but potassa is precipitated with them, and must be removed by repeated boiling with water before the presence of these two elements can be proved by spectrum analysis.

Thallium.—The compounds of this metal give a spectrum with a single intense green line, Tlα, which almost coincides with the δ line of barium. Mere traces of thallium in pyrites may be easily detected by simply heating them in the edge of the flame.

Indium.—The spectrum is characterized by two lines, Inα, in the indigo, and Inβ in the violet. The former is far the most intense, and sufficient for the detection of the metal. To show its presence in sphalerite (zinc blende), the mineral is roasted, decomposed by hydrochloric acid, and the solution diluted and saturated with ammonia. The precipitate containing the oxide of indium is dried and a small portion of it moistened with hydrochloric acid, and placed in the flame with platinum wire. The presence of indium will be indicated by the blue line α.

It is not only those bodies which have the power of giving color to the flame which yield characteristic spectra, for this property belongs to all elementary substances, whether metal or non-metal, solid, liquid, or gas; and it is always noticed when such element is heated to the point at which its vapor becomes luminous, for then each element emits the peculiar light given off by it alone, and the characteristic bright lines become apparent when its spectrum is observed. Most metals require a much higher temperature than the common flame, in order that their vapors should become luminous; but they may be easily heated to the requisite temperature by means of the electric spark, which in passing between two points of the metal in question, volatilizes a small portion, and heats it so intensely as to enable it to give off its peculiar light.

Thus all the metals, — iron, platinum, silver, gold, etc., — may each be recognized by the peculiar bright lines which their spectra exhibit.

The permanent gases also yield characteristic spectra, as hydrogen, nitrogen, oxygen, chlorine, carbonic oxide, etc.

By placing gases, solutions of salts, etc., in glass vessels made for the purpose, and placing them between the illuminating flame and the slit in the tube, so that the light will pass through the gas or liquid, the so-called spectra from absorption are obtained. By this arrangement large portions of the complete spectrum disappear through absorption of corresponding rays of light, or else only dark lines are seen in different parts of the same.

Interesting phenomena of this kind are shown with solutions of certain salts of copper, cobalt, chromium, and permanganate of potassa, and also of the rare metals erbium, didymium, and yttria.

For a description of the great number of spectra which have already been carefully studied and mapped, reference must be made to the large works on this subject recently published and the original memoirs, as it is impossible to give, by a brief description, any idea of the characteristic features of the masses of bright lines constituting these spectra.

By this comparatively new and delicate method of analysis we succeed in obtaining a more accurate knowledge than we have hitherto possessed concerning the chemical composition of terrestrial matter, and also acquire information respecting the chemical nature of the sun, fixed stars, and distant nebulæ, opening out the new sciences of solar and stellar chemistry.

TABLES

SHOWING THE BEHAVIOR OF THE ALKALIES, EARTHS, AND METALLIC OXIDES, ALONE, AND WITH REAGENTS, BEFORE THE BLOW-PIPE.

TABLE I.—BEHAVIOR OF THE ALKALIES AND EARTHS BEFORE THE BLOW-PIPE.

Alkalies.	*Alone on Platinum Wire.*
1. POTASSA. KO. RUBIDIA. RbO. CÆSIA. CsO.	Color the flame violet; but even a minute quantity of soda destroys the reaction.
2. SODA. NaO.	Colors the flame intense reddish-yellow, even in the presence of a large excess of potassa.
3. LITHIA. LiO.	Colors the flame carmine-red, even in the presence of potassa; but soda gives a yellowish-red.
4. AMMONIA. NH_3.	Combined with chlorine, nitric or sulphuric acids, it colors the flame very pale-green.

TABLE I. — Continued.

Alone on Platinum Foil.	
No change.	In solution, change red litmus paper to blue.
No change.	
Turns the foil yellow when fused; but if washed and ignited the color is destroyed, but the foil remains dull.	
No reaction.	Pungent odor, colors red litmus paper blue.

TABLE I.—Continued.

Alkalies.	*On Ch. alone, and in the forceps.*	*With Carbonate of Soda on Ch.*
1. BARYTA. BaO.	The Hydrate fuses, boils, intumesces, and is finally absorbed by the Ch. The Carbonate fuses readily to a transparent glass, which, on cooling, becomes enamel-white. In the forceps it colors the outer flame yellowish-green.	Fuses to a homogeneous mass, which is absorbed by the Ch.
2. STRONTIA. SrO.	The Hydrate behaves like hydrate of Baryta. The Carbonate fuses only on the edges, and swells out in arborescent ramifications which emit a brilliant light, and, when heated with the R. Fl., impart to it a reddish tinge; shows after cooling alkaline reaction. In the forceps, colors the outer flame crimson.	Caustic Strontia is insoluble. The Carbonate, mixed with its own volume of Sd., fuses into a limpid glass, which becomes enamel-white on cooling. At a greater heat the mass boils, and caustic Strontia is formed, which is absorbed by the Ch.
3. LIME. CaO.	Caustic Lime is not changed. The Carbonate loses carbonic acid, becomes whiter and more luminous, and shows after cooling alkaline reaction. In the forceps it colors the outer flame pale-red.	Insoluble. The Sd. passes into the Ch., and leaves the Lime unaltered on its surface.
4. MAGNESIA. MgO.	Undergoes no alteration. The Carbonate becomes caustic and luminous.	It behaves like Lime.
5. ALUMINA. Al^2O^3.	Not changed.	Forms an infusible compound, with slight intumescence. The excess of Sd. is absorbed by the Ch.

TABLE I. — Continued.

With Bx. on Platinum Wire.	*With S. Ph. on Platinum Wire.*
The Carbonate dissolves with effervescence to a limpid glass which, with a certain amount, becomes opaque by flaming; with more, it becomes opaque-white on cooling, even without flaming.	As with Borax.
Same as Baryta.	Same as Baryta.
Readily dissolved to a limpid glass, which becomes opaque by flaming. The Carbonate dissolves with effervescence. On a large addition of Lime the glass becomes cloudy and crystallizes on cooling, but does not become enamel-white, like Baryta or Strontia.	Soluble in large quantities to a limpid glass which, when sufficient Lime is present, becomes opaque by flaming. When saturated, the glass becomes enamel-white on cooling.
It behaves like Lime, but not so crystalline.	Readily soluble to a limpid glass, which becomes opaque by flaming. When saturated, it becomes, on cooling, enamel-white.
Dissolves slowly to a limpid glass, which remains so on cooling, and which cannot be made cloudy by flaming. A large quantity of Alumina makes the glass cloudy and nearly infusible; on cooling, it then assumes a crystalline surface.	Soluble to a limpid glass, which remains clear under all circumstances. If too much Alumina is added, the undissolved portion becomes translucent.

TABLE I. — Continued.

Alkalies.	*On Ch. alone, and in the forceps.*	*With Carbonate of Soda on Ch.*
6. GLUCINA. BeO.	Not changed.	Insoluble.
7. YTTRIA. YO.	Not changed.	Insoluble.
8. ZIRCONIA. Zr^2O^3.	Infusible, but emitting a very glaring light.	Insoluble.
9. ERBIA. EO.	The Oxide becomes lighter-colored and translucent in the R. Fl.	Insoluble.
10. THORIA. ThO.	Not changed.	Insoluble.
11. SILICA. SiO_2.	Not changed.	Soluble, with effervescence to a clear glass.

TABLE I. — Continued.

With Bx. on Platinum Wire.	*With S. Ph. on Platinum Wire.*
Soluble in large quantities to a limpid glass, which becomes opaque by flaming. When Glucina is present in excess, it becomes enamel-white on cooling.	As with Borax.
Like Glucina.	Like Glucina.
Like Glucina.	Dissolves more slowly than with Borax.
Dissolves slowly to a clear glass, which becomes opaque by flaming, or on cooling if in excess.	As with Borax.
In small quantity dissolves to a clear glass, which becomes enamel-white on cooling if in excess; if clear, it cannot be made opaque by flaming.	As with Borax.
Dissolves slowly to a limpid glass, difficultly fusible, and cannot be made opaque by flaming.	Soluble in very small quantities to a limpid glass. The insoluble portion becomes translucent.

TABLE II.—BEHAVIOR OF THE METALLIC OXIDES BEFORE THE BLOW-PIPE.

Metallic Oxides in Alphabetical Order.	*On Charcoal alone.*	*With Carbonate of Soda.*
1. TEROXIDE OF ANTIMONY, OR ANTIMONOUS ACID. SbO^3.	O. Fl.: It is displaced, and deposited upon another part of the Ch. R. Fl.: It is reduced and volatilized. A Ct. of antimonous acid is deposited on the Ch., and a greenish-blue color imparted to the flame.	On Ch. very readily reduced in O. Fl. and R. Fl. The metal fumes and coats the Ch. with oxide of Antimony.
2. ARSENOUS OXIDE. AsO^3.	Volatilizes below a red heat.	On Ch. reduced, with emission of arsenical fumes, which are characterized by a strong garlic odor.
3. TEROXIDE OF BISMUTH. BiO^3.	O. Fl.: On platinum foil it fuses readily to a dark-brown mass, which, on cooling, becomes pale yellow. On Ch. in O. Fl. and R. Fl. reduced to metallic bismuth, which, with long blowing, vaporizes, coating the Ch. with yellow oxide. The Ct., when touched with the R. Fl., disappears without coloring the flame.	Easily reduced to metallic bismuth.
4. OXIDE OF CADMIUM. CdO.	O. Fl.: On platinum foil unchanged. R. Fl.: On Ch. it disappears in a short time, and deposits all over the Ch. a dark-yellow or reddish-brown powder; the color can only be clearly discerned after cooling.	O. Fl.: Insoluble. R. Fl.: On Ch readily reduced; the metal vaporizes and deposits a dark-yellow or reddish-brown Ct. on the Ch. The more remote portion of the coal assumes a variegated tarnish.

TABLE II.—Continued.

With Bx. on Platinum Wire.	*With S. Ph. on Platinum Wire.*
O. Fl.: Dissolves in large quantities to a limpid glass, which, while hot, appears yellowish, but after cooling, colorless. R. Fl.: The glass, when treated only for a short time in the O. Fl., becomes on Ch. grayish and cloudy from particles of reduced antimony. With tin it becomes gray or black, according to the degree of saturation.	O. Fl.: Dissolves with effervescence to a limpid glass, which, while hot, is slightly yellowish. R. Fl.: On Ch. the saturated bead becomes at first cloudy, but afterwards clear again, owing to the volatilization of the reduced antimony. Treated with tin, the glass becomes, after cooling, gray, even if but very little antimonous oxide is present. With strong blowing it becomes clear again.
O	O
O. Fl.: A small quantity is easily dissolved to a clear yellow glass, which, on cooling, becomes colorless. On a large addition of oxide, the glass, while hot, is yellowish-red, becomes yellow on cooling, and when cold is opalescent. R. Fl.: On Ch. the glass becomes at first gray and cloudy, the oxide is reduced to metal with effervescence, and the bead becomes clear again. An addition of tin accelerates the process.	O. Fl.: Readily dissolved to a limpid yellow glass, which, on cooling, becomes colorless. When a greater quantity of oxide is present, the glass may be made enamel-white by flaming, and on a still larger addition it becomes by itself enamel-white on cooling. R. Fl.: On Ch., particularly when tin is added, the glass remains colorless and limpid while hot, but becomes, on cooling, dark-gray and opaque.
O. Fl.: Soluble in large quantity to a limpid yellowish glass, becoming almost colorless on cooling. When highly saturated, it may be made enamel-white by flaming, and when still more oxide is present, it becomes by itself enamel-white on cooling. R. Fl.: Placed on Ch., it enters into ebullition; the oxide is reduced; the reduced metal vaporizes immediately and deposits a dark-yellow Ct.	O. Fl.: Soluble in large quantity to a limpid glass, which, while hot, is yellowish, but colorless when cold; when saturated, it becomes enamel-white on cooling. R. Fl.: On Ch., the oxide becomes slowly and imperfectly reduced. The reduced metal deposits a very feeble Ct. of dark-yellow color. The color is only clearly seen when the mass is cold. An addition of tin facilitates the reduction.

TABLE II. — Continued.

Metallic Oxides in Alphabetical Order.	*On Charcoal alone.*	*With Carbonate of Soda.*
5. SESQUIOXIDE OF CERIUM. Ce^2O^3.	The protoxide is converted into sesquioxide, Ce^2O^3 by the O. Fl., which remains unaltered in the R. Fl.	Insoluble. The Sd. passes into the Ch.; the sesquioxide is reduced to protoxide, which remains on the Ch. as a light-gray powder.
6. SESQUIOXIDE OF CHROMIUM Cr^2O^3.	Not changed in the O. Fl. or R. Fl.	O. Fl.: On platinum wire soluble to a dark yellowish-brown glass, which on cooling becomes opaque and yellow. (Chromic acid.) R. Fl.: The glass becomes opaque and green on cooling. On Ch. it cannot be reduced to metal; the Sd. passes into the Ch., and the oxide remains behind as a green powder, Cr^2O^3.
7. PROTOXIDE OF COBALT. CoO.	O. Fl.: Not changed. R. Fl.: It is reduced to metal, but does not fuse: the mass is attracted by the magnet, and assumes metallic lustre by friction.	O. Fl.: On platinum wire a very small quantity is dissolved to a transparent mass of a pale reddish color, which on cooling becomes gray. R. Fl.: On Ch. reduced to a gray magnetic powder.

TABLE II.—Continued.

With Bx. on Platinum Wire.	*With S. Ph. on Platinum Wire.*
O. Fl.: Soluble to a limpid glass of dark-yellow or red color, which changes on cooling to yellow. When highly saturated with oxide the glass becomes, on cooling, enamel-white. R. Fl.: The yellow glass becomes colorless. A highly saturated bead becomes on cooling enamel-white and crystalline.	O. Fl.: As with Bx., but on cooling, colorless. R. Fl.: Perfectly colorless, hot and cold, thus being distinguished from a sesquioxide of iron glass. Becomes never opaque on cooling, however large the amount of oxide.
O. Fl.: Dissolves but slowly, but colors intensively. If little of the oxide is present, the glass, while hot, is yellow; when cold, yellowish-green; with more oxide it is dark-red while hot, becomes yellow on cooling, and when perfectly cold has a fine yellowish-green color. R. Fl.: The glass is green, hot and cold. The intensity of the color depends on the amount of oxide present. Tin causes no change.	O. Fl.: Soluble to a limpid glass, which, while hot, appears reddish; when cold it has a fine green color. R. Fl.: As in O. Fl.
O. Fl.: Colors very intensively. The glass appears pure smalt-blue, hot and cold. An excess of oxide imparts to the bead a deep bluish-black color. R. Fl.: As in O. Fl.	O. Fl.: As with Bx., but for the same quantity of oxide the color is not quite so deep. R. Fl.: As in O. Fl.

TABLE II. — Continued.

Metallic Oxides in Alphabetical Order.	*On Charcoal alone.*	*With Carbonate of Soda.*
8. OXIDE OF COPPER. CuO.	O. Fl.: Fuses to a black globule, which becomes reduced where it is in contact with the Ch. R. Fl.: Reduced to metal at a temperature below the melting-point of copper. When the heat is increased a globule of metallic copper is obtained.	O. Fl.: On platinum wire soluble to a limpid glass of green color; on cooling it becomes opaque and white. R. Fl.: On Ch. easily reduced to metal, which, when the temperature is sufficiently high, fuses to one or more globules.
9. SESQUIOXIDE OF DIDYMIUM. D^2O^3.	Infusible in the O. Fl. In R. Fl. loses its brown color and becomes gray at a high temperature.	Insoluble. The soda is absorbed by the coal, and the gray oxide remains behind.
10. TEROXIDE OF GOLD. AuO^3.	When heated to ignition it becomes reduced to metal in O. Fl. and R. Fl. The metal fuses easily to a globule.	Does not dissolve in the Sd., but is easily reduced in both flames. The metal fuses readily to a globule. The Sd. passes into the Ch.
11. OXIDE OF INDIUM. InO.	In the O. Fl. becomes dark-yellow when heated, but on cooling is again lighter, and does not fuse. In the R. Fl. is gradually reduced and volatilized, depositing a coat on the coal. A distinct violet flame is produced.	In the O. Fl. insoluble. In the R. Fl. is reduced on coal, and the metal partly volatilizes, coating the coal with oxide, while a portion may be seen as almost silver-white globules in the fused salt.
12. BINOXIDE OF IRIDIUM. IrO^2.	At a red heat becomes reduced; the reduced metal is infusible.	O. Fl.: Does not dissolve in the Sd., but becomes reduced; the metal cannot be fused to a globule. R. Fl: As in O. Fl.

TABLE II. — Continued.

With Bx. on Platinum Wire.	*With S. Ph. on Platinum Wire.*
O. Fl.: A small addition of oxide makes the glass appear green while hot, but blue when cold. A large quantity imparts to it a very deep-green color while hot, becoming greenish-blue when cold. R. Fl.: A glass containing a certain quantity of oxide becomes colorless, but on cooling becomes opaque and red (suboxide). On Ch. the copper may be precipitated in the metallic state, the bead becoming in consequence colorless. A glass containing protoxide, when treated on Ch. with tin, becomes on cooling brownish-red and opaque.	O. Fl.: As with Bx., but for the same amount of oxide the coloration is not so deep. R. Fl.: A glass containing a large quantity of oxide becomes dark-green, which in the moment of refrigeration changes suddenly to brownish-red and opaque. A glass containing but little oxide, when treated on Ch. with tin, appears colorless while hot, but becomes brownish-red and opaque on cooling.
In the O. Fl. soluble to a clear rose-red glass, and remains unaltered in the R. Fl.	Dissolves with more difficulty than in borax, but when strongly saturated is distinctly rose-red.
As with Carbonate of Soda.	As with Carbonate of Soda.
In the O. Fl. dissolves to a clear glass, feebly yellowish while hot, colorless on cooling, and cloudy when much is added. In the R. Fl. the glass is unchanged. On coal the oxide is reduced, volatilizes and coats the coal again with oxide. The flame is violet.	As with Borax, but the glass when treated with tin on coal, becomes gray and cloudy on cooling.
As with Carbonate of Soda.	As with Carbonate of Soda.

TABLE II. — Continued.

Metallic Oxides in Alphabetical Order.	*On Charcoal alone.*	*With Carbonate of Soda.*
13. SESQUIOXIDE OF IRON. Fe^2O^3.	O. Fl.: Not changed. R. Fl: Becomes black and magnetic.	O. Fl.: Insoluble. R. Fl.: On Ch. it is reduced; the mass, when placed in a mortar, pulverized, and repeatedly washed with water to remove the adherent Ch. particles, yields a gray metallic powder which is attracted by the magnet.
14. OXIDE OF LANTHANUM. LaO.	Unchanged.	Insoluble. The soda is absorbed by the coal, leaving the gray oxide behind.
15. OXIDE OF LEAD. PbO.	Minium, when heated on platinum foil, blackens; on increasing the temperature it changes into yellow oxide, which finally fuses to a yellow glass. On Ch. in O. Fl. and R. Fl. almost instantaneously reduced to metal which, with continued blowing, vaporizes, and covers the Ch. with yellow oxide, surrounded by a faint white ring of carbonate. The Ct., when touched with the R. Fl., disappears, imparting to the flame an azure-blue tinge.	O. Fl.: On platinum wire readily dissolved to a limpid glass, which, on cooling, becomes yellowish and opaque. R. Fl.: On Ch. reduced to metal which, with continued blowing, covers the Ch. with oxide.

TABLE II. — Continued.

With Bx. on Platinum Wire.	*With S. Ph. on Platinum Wire.*
O. Fl.: A small amount of oxide causes the glass to look yellow while hot, colorless when cold. When more of the oxide is present the glass, while hot, appears red, and yellow when cold. A still larger quantity makes the glass dark-red while hot, and dark-yellow when cold. R. Fl.: The glass becomes bottle-green. Treated on Ch. with tin it becomes, at first, bottle-green, but afterwards pure vitriol-green.	O. Fl.: When at a certain point of saturation the glass, while hot, appears yellowish-red, and becomes on cooling at first yellow, then greenish, and finally colorless. On a very large addition of oxide it appears, while hot, deep-red, becoming, on cooling, brownish-red, then of a dirty-green color, and finally less brownish-red. R. Fl.: A glass containing but little of the oxide suffers no visible change. When more of the oxide is present it is red while hot, and on cooling becomes at first yellow, then greenish, and finally reddish. Treated with tin on Ch. the glass on cooling becomes at first green, and finally colorless.
In the O. Fl. dissolves to a clear, colorless glass, that becomes enamel-white by flaming when saturated to a certain extent, and when strongly saturated becomes enamel-like of itself on cooling. In R. Fl. the same as in O. Fl.	As with Borax.
O. Fl.: Easily soluble to a limpid yellow glass which, on cooling, becomes colorless. If much oxide is present it may be made cloudy by flaming. A still larger addition of oxide causes the bead to become enamel-yellow on cooling. R. Fl.: The glass diffuses itself over the Ch. and becomes cloudy. With continued blowing the oxide is reduced to metal, with effervescence, and the glass becomes clear again.	O. Fl.: As with Bx. But to obtain a glass which appears yellow while hot, a large addition of the oxide is required. R. Fl.: On Ch. the glass becomes grayish and cloudy. This phenomenon is better observed when tin is added; but the glass can never be made quite opaque. If much of the oxide is present, the Ch. becomes coated.

TABLE II. — Continued.

Metallic Oxides in Alphabetical Order.	*On Charcoal alone.*	*With Carbonate of Soda.*
16. SESQUIOXIDE OF MANGANESE. Mn^2O^3.	O. Fl.: Infusible. When the temperature is sufficiently high, both the sesquioxide and the peroxide are converted into a reddish-brown powder ($MnO + Mn^2O^3$). R. Fl.: The same effect.	O. Fl.: On platinum wire or foil a very small quantity dissolves to a transparent green mass, which on cooling becomes opaque and bluish-green. R. Fl.: On Ch. it cannot be reduced to metal; the Sd. passes into the Ch. and leaves the protoxide behind.
17. PROTOXIDE OF MERCURY. HgO.	Instantly reduced and volatilized.	Heated in a matrass to redness, it is reduced and vaporized. The vapors condense in the neck of the matrass and form a metallic coating, which can be united to a globule by carefully tapping on the matrass.
18. MOLYBDIC ACID. MoO^3.	O. Fl.: Fuses, becomes brown, vaporizes, and deposits on the Ch. a yellow Ct., which nearest to the assay is crystalline. On cooling the Ct. becomes white, and the crystals colorless. Beyond this coat is a thinner non-volatile film of binoxide, which on cooling is dark copper-red, with metallic lustre. R. Fl.: The greater part of the assay is absorbed by the Ch., and may be reduced to metal at a sufficiently high temperature; the metal is in the shape of a gray powder.	O. Fl.: On platinum wire dissolves with effervescence to a limpid glass, which on cooling becomes milk-white. R. Fl.: Fuses with effervescence. The fused mass is absorbed by the Ch., and part of the acid is reduced to metal which may be obtained as a steel-gray powder.

TABLE II. — Continued.

With Bx. on Platinum Wire.	*With S. Ph. on Platinum Wire.*
O. Fl.: Colors very intensively. The glass, while hot, is violet, on cooling it assumes a reddish tinge. When much manganese is added, the glass becomes quite black and opaque; but the color can be seen when the glass, while soft, is flattened with the forceps. R. Fl.: The glass becomes colorless. If the color was very dark, the phenomenon is best observed on Ch. with addition of tin.	O. Fl.: A considerable addition of manganese must be made to produce a colored glass; it then appears, while hot, brownish-violet, and reddish-violet when cold, but never opaque. If the glass contains so small a quantity of manganese that it appears colorless, an addition of nitre will produce the characteristic coloration. A glass containing oxide bubbles and yields gas at a high temperature. R. Fl.: Becomes very soon colorless, and remains then quiet.
O. Fl.: Dissolved in large quantities to a limpid glass which, while hot, appears yellow, but colorless on cooling. A very large amount of acid causes the glass to appear dark-yellow while hot, and opaline when cold. R. F.: A highly saturated bead becomes brown, and opaque when still more acid is present. In a good flame black flocks of MoO^2 separate, and are visible in the yellow glass, if pinched flat.	O. Fl.: Easily soluble to a limpid glass; if but little of the acid is present it is yellowish-green while hot, but when cold almost colorless. On Ch. the glass becomes very dark, and on cooling assumes a beautiful green color. R. Fl.: The glass assumes a very dark, dirty-green color which, on cooling, becomes beautiful bright-green. The same on Ch.; tin deepens the color a little.

TABLE II.— Continued.

Metallic Oxides in Alphabetical Order.	*On Charcoal alone.*	*With Carbonate of Soda.*
19. PROTOXIDE OF NICKEL. NiO.	O. Fl.: Not changed. R. Fl.: On Ch. reduced to metal; the spongy mass cannot be fused to a globule, but assumes metallic lustre by friction; it is attracted by the magnet.	O. Fl.: Insoluble. R. Fl.: Easily reduced to metal, in the shape of bright, white scales, which are attracted by the magnet, if the particles of coal are washed away.
20. NIOBIC ACID. NbO^2.*	In O. Fl. becomes yellowish, but white again on cooling. In R. Fl. the same.	In O. Fl., with equal volume of soda, fuses with effervescence; with more soda, sinks into the coal. In R. Fl. the same. It cannot be reduced to metal.
21. HYPONIOBIC ACID. Nb^2O^3.*	Like Niobic Acid.	Like Molybdic Acid.

* These are the old formulæ, and without doubt the acids are identical, and should have the formula NbO^5.

TABLE II.—Continued.

With Bx. on Platinum Wire.	*With S. Ph. on Platinum Wire.*
O. Fl.: A small quantity colors the bead violet while hot; when cold, pale reddish-brown. More oxide makes the coloration deeper. R. Fl.: The glass becomes gray and cloudy, or even opaque. With continued blowing the minute particles of reduced metal collect together and the glass becomes colorless. This takes place more readily on Ch., especially when tin is added. The nickel then unites with the tin to a globule.	O. Fl.: Soluble to a reddish glass which, on cooling, becomes yellow. A larger addition causes the glass to appear brownish-red while hot, and reddish-yellow when cold. R. Fl.: On platinum wire not changed. On Ch. with tin it becomes, at first, gray and opaque; with continued blowing the nickel becomes reduced, and the glass clear again and colorless.
In O. Fl. dissolves easily to a clear, colorless glass, becoming opaque by flaming with a certain addition, and with more becomes opaque of itself when cool. In R. Fl. a glass which, after treatment in the O. Fl., becomes opaque of itself, on cooling remaining unaltered.	In O. Fl. dissolves largely to a clear glass, yellow while hot, but colorless on cooling. In R. Fl. with a large addition the glass becomes brown. The addition of sulphate iron gives a blood-red bead.
In O. Fl. dissolves easily to a clear glass, which, with the addition of a very large amount, can be made opaque by flaming. In R. Fl. a glass formed in the O. Fl., which opalesces on cooling, becomes clear again. With more it becomes cloudy, and bluish-gray on cooling; with a very large amount it becomes quite opaque and bluish-gray.	The acid prepared from the sesquichloride behaves as follows: In O. Fl. in large quantities it dissolves to a greenish-blue glass, which on continued blowing becomes clear and colorless. In R. Fl. becomes pure blue (on coal pale-brown), and with addition of sulphate of iron blood-red. Prepared from Columbite directly, yields only a brown glass in the inner flame.

TABLE II. — Continued.

Metallic Oxides in Alphabetical Order.	*On Charcoal alone.*	*With Carbonate of Soda.*
22. BINOXIDE OF OSMIUM. OsO^2.	O. Fl.: Converted into osmic acid which, without depositing a Ct., volatilizes with its peculiar pungent odor. R. Fl.: Easily reduced to a dark-brown and infusible metallic powder.	Easily reduced to an infusible metallic powder, which may be obtained pure by washing away the coal.
23. PROTOXIDE OF PALLADIUM. PdO.	Reduced at a red heat; but the metallic particles are infusible.	Insoluble. The Sd. passes into the Ch., and leaves the Palladium behind as an infusible powder.
24. BINOXIDE OF PLATINUM. PtO^2.	Like Palladium.	Like Palladium.
25. SESQUIOXIDE OF RHODIUM. R^2O^3.	Like Palladium.	Like Palladium.
26. SESQUIOXIDE OF RUTHENIUM. Ru^2O^3.	Like Palladium.	Like Palladium.
27. OXIDE OF SILVER. AgO.	Easily reduced to metallic silver, which unites to one or more globules.	Instantly reduced. The Sd. passes into the Ch., and the metal unites to one or more globules.
28. TANTALIC ACID. TaO^5.	In O. Fl. becomes slightly yellow, but is white again when cold. In R. Fl. the same.	In R. Fl. with more than an equal volume of soda fuses on coal to a bead with effervescence, and soon spreads out; with more soda sinks into the coal. In R. Fl. the same. It cannot be reduced to metal.

TABLE II. — Continued.

With Bx. on Platinum Wire.	*With S. Ph. on Platinum Wire.*
O. Fl. and R. Fl.: Reduced, but not dissolved; the metallic particles cannot be fused to a globule.	As with Bx.
Like Palladium.	Like Palladium.
Like Palladium.	Like Palladium.
Like Palladium.	Like Palladium.
O. Fl.: In part dissolved, and in part reduced. On cooling, the glass becomes opalescent or milk-white, according to the amount of oxide present. R. Fl.: The glass at first becomes gray, but afterwards limpid and colorless.	O. Fl.: Imparts to the bead a yellowish color. When much of the oxide is present, the glass, when cold, is opalescent, and appears yellowish at daylight, reddish at candle-light. R. Fl.: As with Bx.
In O. Fl. dissolves easily to a clear glass, which with a certain amount appears yellowish while hot, colorless on cooling, and can be made opaque by flaming. With still more the glass becomes enamel-white of itself on cooling. In R. Fl. same as in O. Fl.	In O. Fl. dissolves largely to a clear glass, which with a very large amount is yellowish while hot, but colorless on cooling. In R. Fl. the above glass is unchanged.

TABLE II. — Continued.

Metallic Oxides in Alphabetical Order.	*On Charcoal alone.*	*With Carbonate of Soda.*
29. TELLUROUS ACID. TeO^2.	O. Fl.: Fuses, and is reduced with effervescence. The reduced metal becomes instantly vaporized and covers the Ch. with tellurous acid; the Ct. usually has a red or dark-yellow edge. R. Fl.: As in O. Fl.; the outer flame appears of a bluish-green color.	Soluble, on platinum wire, to a limpid and colorless glass, which on cooling becomes white. On Ch. reduced and volatilized, depositing a Ct. of tellurous acid.
30. BINOXIDE OF TIN. SnO^2.	O. Fl.: The protoxide burns, like tinder, to binoxide. The binoxide becomes very luminous and appears, while hot, yellowish, but assumes on cooling a dirty-white color. R. Fl.: With a powerful and continued flame it may be reduced to metal, a trifling of SnO^2 being formed near the metal.	O. Fl.: On platinum wire it forms with Sd., with effervescence, an infusible compound. R. Fl.: On Ch. reduced to metallic tin.
31. TITANIC ACID. TiO^2.	O. Fl.: Assumes, on heating, a yellow color, and becomes white again on cooling. Suffers no other change. R. Fl.: As in O. Fl.	O. Fl.: On Ch. it dissolves, with effervescence, to a dark-yellow glass, which, on cooling, crystallizes, and thereby evolves so much heat that the globule glows again strongly. When cold it is grayish-white to white. R. Fl.: As in O. Fl.; cannot be reduced to metal.

TABLE II. — Continued.

With Bx. on Platinum Wire.	*With S. Ph. on Platinum Wire.*
O. Fl.: Soluble to a limpid and colorless glass which, on Ch., becomes gray from reduced metal. R. Fl.: On Ch. becomes at first gray, afterward colorless. The Ch. becomes coated with tellurous acid.	As with Borax.
O. Fl.: A very small quantity dissolves slowly to a limpid and colorless glass, which remains so on cooling, and not becoming opaque by flaming. R. Fl.: From a highly saturated glass a part of the oxide may be reduced on Ch.	O. Fl.: As with Borax. R. Fl.; The glass, containing oxide, suffers no change on coal or platinum wire.
O. Fl.: Easily soluble to a limpid glass which, when containing a large quantity, appears yellow while hot, but becomes colorless on cooling. When containing a very large quantity it is enamel-white when cold. R. Fl.: When containing but little titanic acid the glass becomes yellow; when more, dark-yellow to brown. A saturated glass becomes enamel-blue by flaming.	O. Fl.: Easily dissolved to a limpid glass which, when containing a large quantity, appears yellow while hot, but becomes colorless on cooling. R. Fl.: Appears yellow while hot, but, on cooling, reddens and finally assumes a violet color. If iron is present the glass, on cooling, becomes brownish-red; with tin, or if the amount of titanium is very small, metallic zinc, on Ch. the glass becomes violet, unless the amount of iron be very considerable.

TABLE II. — Continued.

Metallic Oxides in Alphabetical Order.	*On Charcoal alone.*	*With Carbonate of Soda.*
32. TUNGSTIC ACID. WO^3.	O. Fl.: Not changed; at a very high temperature converted into oxide. R. Fl.: Blackens, being converted into oxide, but does not fuse.	O. Fl.: On platinum wire it dissolves to a limpid and deep-yellow glass, which, on cooling, becomes crystalline and opaque, and of white or yellowish color. R. Fl.: With very little Sd. on Ch. it is reduced to metal; with more Sd. it forms a yellow compound of metallic lustre which passes into the Ch.
33. SESQUIOXIDE OF URANIUM. U^2O^3.	O. Fl.: Infusible; but assumes a dirty yellowish-green color. R. Fl.: Blackens, owing to the formation of protoxide.	O. Fl.: Insoluble. With a certain amount of Sd. the mass becomes yellowish-brown, and with more passes into the Ch. R. Fl.: As in O. Fl.; no reduction to metal takes place.
34. VANADIC ACID. VO^3.	Fusible. Where it is in contact with the Ch. it becomes reduced and passes into the Ch. The rest assumes the lustre and color of graphite.	Unites to a fusible mass which is absorbed by the Ch.

TABLE II. — Continued.

With Bx. on Platinum Wire.	*With S. Ph. on Platinum Wire.*
O. Fl.: Like titanic acid. R. Fl.: A glass, containing but little tungstic acid, is not changed. When more, it becomes yellow, and, on cooling, yellowish-brown. On Ch. the same reaction is produced with a less saturated bead. Tin deepens the colors.	O. Fl.: Easily dissolved to a limpid and colorless bead, which, when highly saturated, appears yellow while hot. R. Fl.: With little blowing the glass appears, while hot, of a dirty green color, blue on cooling; with strong blowing it becomes, on cooling, bluish-green. On Ch. with tin, deep green. If iron is present the glass, on cooling, becomes brownish-red; with tin on Ch. the glass becomes blue or, if the amount of iron is considerable, green.
O. Fl.: Behaves like sesquioxide of iron. When highly saturated the glass may be made enamel-yellow by flaming. R. Fl.: Behaves like sesquioxide of iron. The green bead, when at a certain point of saturation, may be made black by flaming. On Ch. with tin it becomes dark-green.	O. Fl.: Dissolves to a limpid yellow glass, which, on cooling, becomes yellowish-green. R. Fl.: The glass assumes a dirty green color which, on cooling, changes to a fine green. With tin on Ch. the color deepens.
O. Fl.: Dissolved to a limpid glass which, when the quantity of vanadic acid is small, appears colorless, when larger yellow, and which, on cooling, becomes greenish. R. Fl.: The glass, while hot, appears brownish, and assumes a fine green color on cooling.	O. Fl.: Soluble to a limpid glass which, if sufficient vanadic acid is present, appears dark-yellow while hot, and becomes light-yellow on cooling. R. Fl.: As with Borax.

TABLE II. — Continued.

Metallic Oxides in Alphabetical Order.	*On Charcoal alone.*	*With Carbonate of Soda.*
35. OXIDE OF ZINC. ZnO.	O. Fl.: When heated becomes yellow, and, on cooling, white again. It fuses not, but becomes very luminous. R. Fl.: Is slowly reduced; the reduced metal becomes rapidly re-oxidized and the oxide deposited on another place of the Ch., being yellowish while hot, and white on cooling.	O. Fl.: Insoluble. R. Fl.: On Ch. it becomes reduced. The metal vaporizes and coats the Ch. with oxide. With a powerful flame the characteristic zinc flame is sometimes produced.

TABLE II. — Continued.

With Bx. on Platinum Wire.	*With S. Ph. on Platinum Wire.*
O. Fl.: Dissolves readily, and in large quantity, to a limpid glass, which appears yellowish while hot; on cooling it is colorless. When much of the oxide is present, the glass may be made enamel-white by flaming; and on a still larger addition it becomes enamel-white on cooling. R. Fl.: The saturated glass becomes at first gray and cloudy, and finally transparent again. On Ch. the oxide becomes reduced, the metal vaporizes and coats the Ch. with oxide.	As with Borax.

TABLE III.—THE METALLIC OXIDES ARRANGED WITH REFERENCE TO THE COLORS WHICH THEY IMPART TO THE FLUXES.

WITH BORAX IN THE OXIDIZING FLAME PRODUCE:

a.—Colorless Beads.

HOT AND COLD.	Silica, Alumina, Binoxide of Tin, Baryta, Strontia, Lime, Magnesia, Glucina, Yttria, Zirconia, Thoria, Oxides of Lanthanum and Silver, Tantalic, Niobic, Tellurous Acids;	When highly saturated opaque (white) by flaming.
	Titanic, Tungstic, Molybdic Acids; Oxides of Zinc, Cadmium, Lead, Bismuth, and Antimony;	When feebly saturated.

b.—Yellow Beads.

HOT.	Titanic, Tungstic, and Molybdic Acids; Oxides of Zinc and Cadmium;	When highly saturated; on cooling, colorless, and cloudy by flaming.
	Oxides of Lead, Bismuth, and Antimony;	When highly saturated; on cooling, colorless.
	Sesquioxides of Cerium, Iron, and Uranium;	When feebly saturated; on cooling, colorless.
	Sesquioxide of Chromium, when fully saturated; when cold, yellowish-green.	
	Vanadic Acid; when cold, greenish-yellow.	

c.—Red to Brown Beads.

HOT.	Sesquioxide of Didymium, rose-red; same when cold.
	Sesquioxide of Cerium; on cooling, yellow; enamel-like by flaming.
	Sesquioxide of Iron; on cooling, yellow.
	Sesquioxide of Uranium; on cooling, yellow; enamel-yellow by flaming.
	Sesquioxide of Chromium; on cooling, yellowish-green.
	Sesquioxide of Iron, containing Manganese; on cooling, yellowish-red.
COLD.	Protoxide of Nickel, reddish-brown to brown; violet while hot.
	Sesquioxide of Manganese, violet-red; violet while hot.
	Protoxide of Nickel, containing Cobalt; violet while hot.

TABLE III.—Continued.

WITH BORAX IN THE REDUCING FLAME PRODUCE:

a. — *Colorless Beads.*

HOT AND COLD.	Silica, Alumina, Binoxide of Tin.	
	Baryta, Strontia, Lime, Magnesia, Glucina, Yttria, Zirconia, Thoria; Oxides of Lanthanum and Cerium, Tantalic Acid;	When highly saturated become cloudy or opaque by flaming.
	Sesquioxide of Manganese and Oxide of Indium; sometimes the Manganese glass, on cooling, pale rose-colored.	
	Niobic and Hyponiobic Acids; when feebly saturated.	
	Oxides of Silver, Zinc, Cadmium, Lead, Bismuth, Antimony, Nickel, Tellurous Acid;	With strong blowing; with feeble blowing, gray.
HOT.	Oxide of Copper; when highly saturated, on cooling, opaque and red.	

b. —*Yellow to Brown Beads.*

HOT.	Titanic Acid, yellow to brown; when highly saturated, enamel-blue by flaming.
	Tungstic Acid, yellow to dark-yellow; when cold, brownish.
	Molybdic Acid, brown to opaque. If the bead is flattened black binoxide of Molybdenum may be noticed.
	Vanadic Acid, brownish; green when cold.

c. — *Blue Beads.*

HOT.	Protoxide of Cobalt; retains its color on cooling.

TABLE III. — Continued.

WITH BORAX IN THE OXIDIZING FLAME PRODUCE:

d. — Violet Beads (amethyst-colored).

HOT.	Protoxide of Nickel; on cooling, reddish-brown to brown. Sesquioxide of Manganese; on cooling, violet-red. Protoxide of Nickel, containing Cobalt; on cooling, brownish. With excess of Cobalt, amethyst color when cold. Protoxide of Cobalt containing Manganese; also when cold.

e. — Blue Beads.

HOT.	Protoxide of Cobalt; retains its color on cooling.
COLD.	Oxide of Copper, when highly saturated, greenish-blue; green while hot.

f. — Green Beads.

HOT.	Oxide of Copper; when cold, blue or greenish-blue.	
	Sesquioxide of Iron, containing Cobalt or Copper. Oxide of Copper, containing Iron or Nickel.	On cooling, the color changes, according to the proportion in which the various oxides are present, to light-green, blue, or yellow.
COLD.	Sesquioxide of Chromium, yellowish-green; yellow to red while hot.	

TABLE III. — Continued.

With Borax in the Reducing Flame produce:

d.— Green Beads.

Hot and Cold.	Sesquioxide of Iron, yellowish-green; especially when cold. Sesquioxide of Uranium, yellowish-green; when highly saturated becomes black by flaming. Sesquioxide of Chromium, light to dark emerald-green.
Hot.	Vanadic Acid, chrome-green; brownish while hot.

e.— Gray and Cloudy Beads.

Cold.	Oxides of Silver, Zinc, Cadmium, Lead, Bismuth, Antimony, Nickel, Tellurous Acid;	With feeble blast; with strong blast, colorless.
	Niobic and Hyponiobic Acids; when highly saturated.	

f.— Red Beads.

Cold.	Oxide of Copper, opaque, when highly saturated; colorless while hot. Sesquioxide of Didymium, rose-red.

TABLE III. — Continued.

WITH SALT OF PHOSPHORUS IN OXIDIZING FLAME PRODUCE:

a. — Colorless Beads.

HOT AND COLD.	Silicic Acid; soluble only in minute quantity.	
	Alumina, Binoxide of Tin; soluble with difficulty.	
	Baryta, Strontia, Lime, Magnesia, Glucina, Yttria, Zirconia, Thoria, Oxide of Lanthanum, Tellurous Acid;	When highly saturated become white and opaque by flaming.
	Tantalic, Niobic, Titanic, Tungstic Acids; Oxides of Zinc, Cadmium, Indium, Lead, Antimony, and Bismuth;	If not too highly saturated.

b. — Yellow Beads.

HOT.	Tantalic, Niobic, Hyponiobic, Titanic, Tungstic Acids; Oxides of Zinc, Cadmium, Indium, Lead, Antimony, and Bismuth;	When highly saturated; colorless on cooling.
	Oxide of Silver, yellowish; when cold, opalescent.	
	Sesquioxide of Iron; Sesquioxide of Cerium;	When feebly saturated; on cooling, colorless. If highly saturated, red while hot, yellow when cold.
	Sesquioxide of Uranium; when cold, yellowish-green.	
	Vanadic Acid, deep-yellow; when cold, of a lighter shade.	
COLD.	Protoxide of Nickel; while hot, reddish.	

c. — Red Beads.

HOT.	Sesquioxide of Iron; Sesquioxide of Cerium;	When highly saturated; when cold, yellow.
	Sesquioxide of Didymium, rose-colored when highly saturated.	
	Protoxide of Nickel, reddish; when cold, yellow.	
	Sesquioxide of Chromium, reddish; when cold, emerald-green.	

TABLE III. — Continued.

WITH SALT OF PHOSPHORUS IN REDUCING FLAME PRODUCE:

a.—Colorless Beads.

	Substances	Remarks
HOT AND COLD.	Silica, but slightly soluble.	
	Alumina, Binoxide of Tin, soluble with difficulty.	
	Baryta, Strontia, Lime, Magnesia, Glucina, Yttria, Zirconia, Thoria, Oxide of Lanthanum;	When highly saturated become opaque by flaming.
	Sesquioxides of Didymium, Cerium, Manganese.	
	Oxides of Silver, Zinc, Cadmium, Lead, Bismuth; Tantalic Acid; Oxides of Silver, Cadmium, Zinc, Indium, Bismuth, Lead, and Antimony; Tellurous Acid; Protoxide of Nickel, on Charcoal;	With continued blowing; otherwise gray.

b.—Yellow to Red Beads.

	Substances	Remarks
HOT.	Sesquioxide of Iron; on cooling, greenish, then reddish.	
	Titanic Acid, yellow; on cooling, violet.	
	Vanadic Acid, brownish; when cold, emerald-green.	
	Titanic Acid containing Iron; Tungstic Acid containing Iron;	Yellow; when cold, blood-red.
	Niobic Acid containing Iron;	Brownish-red; when cold, deep-yellow.

c.—Violet (amethyst) Beads.

	Substances
COLD.	Niobic Acid, when highly saturated; while hot, of a pale dirty-blue color.
	Titanic Acid; yellow while hot.

TABLE III. — Continued.

WITH SALT OF PHOSPHORUS IN OXIDIZING FLAME PRODUCE:

d.—Violet (amethyst) Beads.

HOT.	Sesquioxide of Manganese, brownish-violet; on cooling, pale reddish-violet. Oxide of Didymium; when cold, of a lighter shade.

e.— Blue Beads.

HOT.	Protoxide of Cobalt; when cold, of the same color.
COLD.	Oxide of Copper; green while hot.

f.—Green Beads.

HOT.	Sesquioxide of Iron, containing Cobalt or Copper; Oxide of Copper, containing Iron or Nickel; — On cooling, the color changes, according to the proportion in which the various oxides are present, to light-green, blue, or yellow. Oxide of Copper; when cold, blue or greenish-blue if highly saturated. Molybdic Acid, yellowish green; when cold, of a lighter shade.
COLD.	Sesquioxide of Uranium, yellowish-green; while hot, yellow. Sesquioxide of Chromium, emerald-green; while hot, reddish.

TABLE III. — Continued.

With Salt of Phosphorus in Reducing Flame produce:

d. — *Blue Beads.*

Cold. {
Protoxide of Cobalt; of the same color when hot.
Tungstic Acid; while hot, brownish.
Niobic Acid, when very highly saturated; while hot, of a dirty-blue color.

e. — *Green Beads.*

Cold. {
Sesquioxide of Uranium; while hot, less bright.
Molybdic Acid; while hot, of a dirty-green color.
Vanadic Acid; while hot, brownish.
Sesquioxide of Chromium; while hot, reddish.

f. — *Gray and Cloudy Beads.*

Cold. { Oxides of Silver, Zinc, Cadmium, Indium, Lead, Bismuth, Antimony, Nickel; Tellurous Acid. } Takes place quickest with Tin on Ch.; with continued blowing, colorless.

g. — *Red Beads.*

Cold. {
Oxide of Copper, opaque, when highly saturated, or with Tin on Charcoal.
Sesquioxide of Didymium, rose-colored.

APPENDIX.

The following piece of apparatus is very convenient for determining the specific gravity of minerals.*

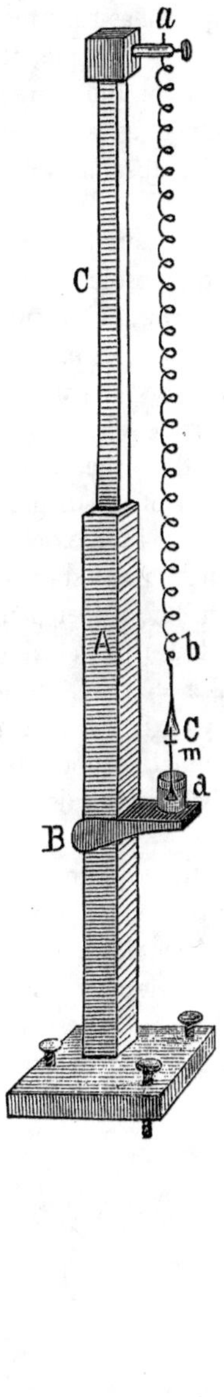

JOLLY'S SPRING BALANCE.

A WIRE wound in a spiral form is suspended at *a*, and has attached at its lower end *b* two pans, *c* and *d*. The pan *d* dips into water. The vessel containing the water, in which the pan is suspended, is placed upon a shelf, which may be moved up or down on the standard of the balance. A mark at *m* shows the extension of the spiral on the mirror scale, which is also attached to the standard. In reading, the mark is made to cover the reflection on the mirror.

If weights increasing a tenth of a gram are added successively to the pan in the air, it is found that the extension of the wire is in proportion to the weight added. Conically wound wire, with its greatest diameter at *a*, and its smallest at *b*, shows precisely the relation between the size of the load and the extension of the spiral.

The manner of using the balance is extremely simple. Before the substance is placed on the pan, the place of the mark is observed on the scale. A known weight is then placed on the upper pan,

* The balance is manufactured by Geo. Wale & Co., at the Hoboken Institute.

and the shelf B, moved down as far as necessary—so far that with the consequent extension of the spiral, the pan *d* will again sink into the water, when a second reading is made. The difference in these figures give the number of divisions on the scale over which the spiral has been made to pass by the weight. If it is found, for example, that with a weight of 1 gram the extension of the spiral is 122.2 divisions, while with some substance, as a piece of mineral, the extension is only 54.4, the absolute weight of the substance will be $\frac{54.4}{122.2}=0.445$.

If, however, the absolute weight is not desired, only the specific, it is not necessary to express the absolute weight in grams. Three readings are made: first, with empty pan; second, with substance placed on the upper pan; and the third after placing the same substance on the pan under water. The difference between the first two gives the absolute weight expressed in divisions of the scale, and the difference in the last two gives the weight of the displaced water. The quotient of these differences is therefore the specific weight. If the mark with empty pans stands at 64.2, and with the substance placed in the upper pan at 275.3, and with the same substance in lower pan at 220.8, then the absolute weight is $275.3-64.2=211.1$, and loss of weight in water is $275.3-220.8=54.5$. Specific weight will be $\frac{211.1}{54.5}=3.85$. The second decimal is not always certain, but by proper arrangement of the spiral is found as reliable as the ordinary balance.

If the specific weight of a fluid is to be determined, both pans are taken off, and in place of them a glass of about 1 c.c. in size is suspended by a fine platinum wire. The loss of weight in water and in other liquids is shown by the scale as in the former case.

As shown in the drawing, the shelf B is attached to the

standard A. The movable standard C has the same length as A, can be raised or lowered, and made fast at any point. C is drawn out, according to the length of the spiral, so far that the mark with the pans empty stands opposite one of the upper divisions of the scale; which, to show the whole extension of the spiral, must be at least 600 m.m. long.

Every spiral at first shows a little elasticity, which grows less, and which during any one experiment amounts to really nothing.

In order to test the fusibility of a mineral, a small splinter, having a sharp edge or point, should be broken off and held in the forceps at a short distance beyond the point of the inner blue flame, so that the sharp edge is strongly heated. If a gas flame be employed, the mineral must be held somewhat further from the point of the blue flame than is necessary in the case of an oil-lamp, in order to prevent any reduction taking place, which would materially interfere with the results. If a powdered substance is to be tested, or one which decrepitates when heated, and which must therefore be previously pulverized, the following process may be resorted to. A small quantity of the powder is made into a paste with water, and spread upon a piece of charcoal; it is then dried, and strongly heated with an oxidizing flame; it will then (generally) cohere sufficiently to allow of its being taken up between the forceps and tested in the usual manner. Care must be taken that the substance, if a fusible one, and one which acts upon platinum, does not fuse upon the platinum points of the forceps.

INDEX.

THE END.

www.ingramcontent.com/pod-product-compliance
Lightning Source LLC
LaVergne TN
LVHW020232110826
845151LV00003B/903

9781425530457